# A New Geometry of Musical Chords in Interval Representation: Dissonance, Enrichment, Degeneracy and Complementation

Miguel Gutierrez, Makoto Taniguchi

A New Geometry of Musical Chords in Interval Representation:
Dissonance, Enrichment, Degeneracy and Complementation

iUniverse books may be ordered through booksellers or by contacting:

iUniverse
1663 Liberty Drive
Bloomington, IN 47403
www.iuniverse.com
1-800-Authors (1-800-288-4677)

ISBN: 978-1-4502-2797-1 (sc)
ISBN: 978-1-4502-2798-8 (e-book)

Printed in the United States of America

iUniverse rev. date: 6/15/2010

To Makoto Taniguchi  (1971-2008)


In memory of my dear friend whose enthusiasm and
tireless labor made this project possible.
His untimely death left Scriabin, Sofronitsky and the
Tetrahedron without, perhaps, their most ardent admirer.

# Preface

The aim of this monograph is to present a self-contained and fairly elementary mathematical approach to the theory of musical chords. Hopefully it will add some light to the growing interest of mathematicians trying to explore the borderland territory between music and mathematics and, I think, even greater the population of music theorists with great affinity for the mathematical language.

The contents of this book are totally independent from the current music theory literature, and this fact is clearly reflected in the scarce body of bibliography quoted in the reference section. Needless to say those were the only sources consulted in the confection of this monograph.

In this sense I need hardly to add that my acquaintance with classical music theory is simply that of an intruder and nothing could be further from my intentions than to pose as an authority on questions of conventional harmony. However this book introduces what I believe to be a fresh, elegant and highly original view at some aspects of musical chord theory.

The main idea in this project originated with the strong belief that the most advantageous way of displaying music chord types, i.e., chords independent of their specific note content, is by looking at the intervallic distances between adjacent notes. For every chord the sum of these distances equals twelve. This is in fact a discrete barycentric condition which leads itself to a geometrical representation of the chordal space as a simplicial grid. Chords appear as points in the grid and their musical inversions would generate beautiful polyhedra inscribe in concentric spheres centered at the barycenter. The radii of these spheres would effectively quantify the evenness and thus the consonance of the chords. Section 1 introduces this geometric model together with a convenient metric in a very straight forward manner.

Section 2 elaborates on different types of symmetries inherent in the model. The inversion symmetry and a new symmetry, Y-symmetry (whose name was inspired by Forte's Z-symmetry) relates chords that are geometrically congruent but are not related by inversion.

In Section 3 we introduce two morphisms: enrichments and reductions that allow us to navigate smoothly through different chord cardinalities.

In Section 4 the geometry really sparkles. This is the most significant part of the work. Due to internal symmetries chords collapse into lower dimension structures, given rise to what we call degenerated chords. The usage of circulant matrices and their eigenvalues proves to be very fruitful.

Finally Section 5 treats complementation in interval form through the introduction of a very unique idempotent map. Characterizations in terms of the morphisms introduced earlier are thoroughly discussed.

Musical oriented readers looking for applicable results won't be disappointed. Some of these results are implicit in the theory, others are clearly stated (for instance the example at the end of Section 3) and others might have escaped the scrutiny of the authors' insight being trapped in some mathematical technicalities. However there is no doubt in my mind that a keen reader will spot many instances where meaningful musical applications are suggested.

Finally a few words about Makoto Taniguchi, the co-author of this monograph. His untimely passing left a painful vacuum in the lives of his family and friends. He was the encouraging voice in the beginning of this endeavor and eventually became trapped in what he coined "the beauty of the four note case". He was responsible for every single diagram and table in the manuscript. He even created what we called "the toy", a beautiful computer program on Excel displaying every single four note chord from any chosen angle. His skills were remarkable.

I extend my gratitude to the editorial team and all those involved in the publication of this monograph at iuniverse.


Miguel Gutierrez

# A New Geometry of Musical Chords in Interval Representation: Dissonance, Enrichment, Degeneracy and Complementation.

Miguel Gutierrez, Makoto Taniguchi

**Contents**

# 0. Introduction

Despite the fact that musical consonance and its antonym, dissonance have experienced attempts of rigorous measurements since ancient times, the musical literature is saturated with vague and subjective terms, like "pleasantness" and "unpleasantness" to characterize these concepts. The perception of consonance has, not only personal but cultural connotations: the "slendro" scale, which partitions the octave in five equal intervals, commonly used in vocal and instrumental music in Bali and Java would sound unpleasantly dissonant to most western ears. Numerous efforts, all welcome, to quantify consonance-dissonance appeared throughout our history; however failures and drawbacks inherent in these models prevent them from global recognition and adoption.

The Pythagorean order of dissonance by the increasing size of the numbers involved in the frequency ratios appeared in the $5^{th}$ century BC and survived, with minor modifications by Boethius, Guido D'Arezzo, Ramos de Pareja, and Zarlino among others, through the $16^{th}$ century and beyond. This theory fails to consider consonant a well tempered $5^{th}$ (virtually indistinguishable from a pure $5^{th}$) since its frequency ratio will be the irrational $1:2^{7/12}$.

Helmholtz, with his "Theory of Beats", classifies intervals as very dissonant if the number of beats produced by the two notes in the interval approaches 33 per second, and consonant if they are less than 6 or more than 120 per second. One of the drawbacks of this theory is that the dissonance in the interval changes dramatically with the octave where the interval lies in: c-e 33 beats, c'-e' 66 beats and c"-e" 132 beats.

The "Fusion Theory" of Carl Stumpf offers a psychological explanation based on large population experimentation. The listener is asked whether he or she perceives a musical interval as one or two notes. The problem with this theory lies with the disparity of the outcomes and the highly subjective component in different experiments (Tenney [9]).

All these theories deal with consonance-dissonance in musical intervals. Musical chords, even to this day, are considered consonant or dissonant if they include consonant or dissonant intervals. More recently, experiments analyzing spectral components of chordal tones with respect to critical bands, performed by Pomp and Levelt, Kameoka and Kuriyagawa, and Huron and Sellmer ([6],[4],[3]) , lead to the assertion that dissonance-consonance strongly depends on the evenness of the tone chordal distribution and the proximity of the notes within the chord. It is well accepted nowadays that the highest consonance in musical chords occurs when the notes in the chord partition the octave in equal or almost equal intervals. At the same time, chords that contain notes that are clustered close together are considered very dissonant.

In view of this principle we are aiming at the construction of a geometric model with a convenient distance (metric space) that effectively quantifies the evenness of the chordal-tone spacing for every single chord type.

In our treatment we want to classify chord types, i.e., chords independent of their specific note content, by looking at the intervallic distances between consecutive notes. For instance, in the classic pitch-class notation, [0, 4, 7] represents a C major chord, [2, 6, 9] a D major chord, and [9, 1, 6] the representation of an A major chord.

Notice that the intervallic distances obtained by subtracting adjacent note values 4, 3, and 5 are the same for all three cases of major triads mentioned above. Thus (4, 3, 5) in our notation is the intervallic representation of any major chord. Notice furthermore that the sum of the intervals of this chord ( in fact any chord) equals 12.

This further abstraction of the representation of a musical object proves to be ideal for the analysis of chord types. Some of its advantageous properties are bulleted bellow:

- Since the entries of the n-tuples are interval distances, octave shift, transposition and cardinality change are intrinsically disregarded.

- We gain a dimension in the visualization of musical chords. Four-note chords and their cyclic orbits appear as tetrahedrons in $\mathbb{R}^3$.

- In pitch-class notation [0,4,7] , [0,3,8] and [0,5,9] are the normalized representatives of cyclic permutations ( inversions in traditional harmony) of a major chord. In interval notation these permutation appear as natural rotations of intervals: (4,3,5), (3,5,4) and (5,4,3).

- Barycentric chords with maximal even tonal distribution appear as $B = \left( \dfrac{12}{n}, \dfrac{12}{n}, ..., \dfrac{12}{n} \right)$ ,i.e., (4,4,4) in the case of the triads and (3,3,3,3) in the four note case.

- The inversion of a chord is formed by displaying the "retrograde" representation of the original chord. Thus, the inversion of (1,2,3,6) will be (6,3,2,1).

- The chordal "prime form" in the Forte (Rham-Morris) sense is formed by simply ordering the intervals in decreasing order from right to left. So (1,3,2,2,4) will be the prime form of (2,2,4,1,3).
- The internal symmetries of the chord appear strikingly obvious when notated in interval form. For instance (1,5,1,5) and (1,1,5,5) are the interval representatives of 4-9=[0,1,6,7] and 4-6=[0,1,2,7] respectively.( see figures 11 and 12 ahead).

Moreover the algebraic relation $\sum_{i=1}^{n} x_i = 12$ , where the $x_i$ are positive integers, is a discrete barycentric condition and consequently represents a discrete simplicial ( hyper- tetrahedral ) grid $\mathcal{T}^n$ centered at its barycenter $B$.

In classical geometry the set of points in the Euclidian space $\mathbb{R}^3$ verifying the above condition is represented by the triangular surface in Figure 2.

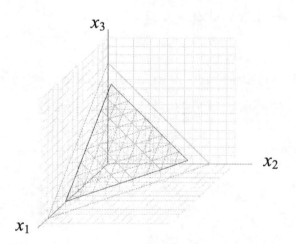

**Figure 2** The surface $x_1 + x_2 + x_3 = 12$ in Euclidian space $\mathbb{R}^3$

In displaying chords in interval representation, the coordinate entries are restricted to be natural numbers constrained to sum to 12. In Euclidian space $\mathbb{R}^3$, these points are coplanar and form a triangular grid. The grid lines on the triangular surface in $\mathbb{R}^3$ correspond to constant values of each variable along each of the grid lines, e.g. $x_1 = 1$, $x_1 = 2$, etc. This leads directly to the use of a triangular grid in $\mathbb{R}^2$, that is, the use of a discrete barycentric coordinate system $(x_1, x_2, x_3)$ as displayed in Figure 3a.

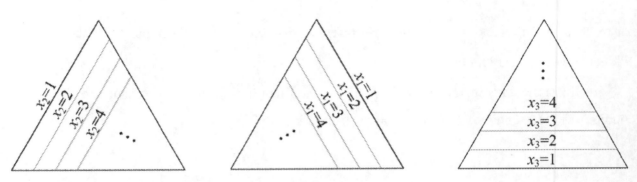

**Figure 3a** Barycentric coordinate system

Following the same procedure used in plotting points in a Cartesian coordinate system, a chord $(x_1, x_2, x_3) = (3, 2, 7)$ in our model is the point where the triangular grid lines $x_1 = 3$, $x_2 = 2$, and $x_3 = 7$ intersect. (See Figure 3b.)

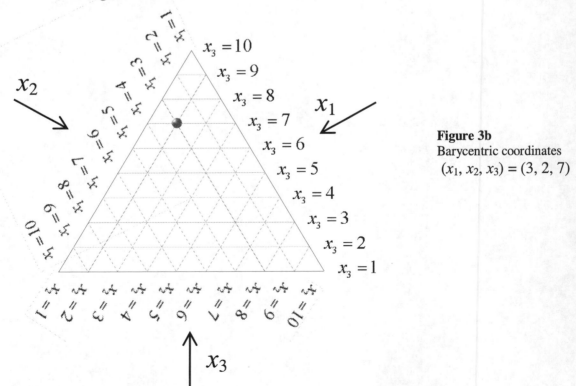

**Figure 3b**
Barycentric coordinates
$(x_1, x_2, x_3) = (3, 2, 7)$

4

The representation of 4-note chords can be extended in a similar manner using the algebraic condition $x_1 + x_2 + x_3 + x_4 = 12$. The set of points in the Euclidian space $\mathbb{R}^3$ satisfying the above constraint is represented by parallel triangular surfaces (Figure 4) of the form

$$x_1 + x_2 + x_3 = 3, \quad \text{where } x_4 = 9$$
$$x_1 + x_2 + x_3 = 4, \quad \text{where } x_4 = 8$$
$$x_1 + x_2 + x_3 = 5, \quad \text{where } x_4 = 7$$
$$\vdots$$
$$x_1 + x_2 + x_3 = 11, \quad \text{where } x_4 = 1$$

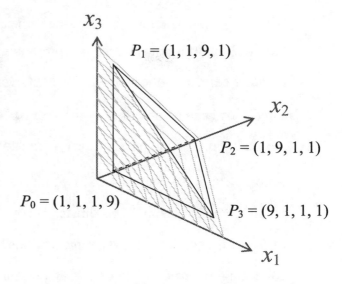

**Figure 4** The surface $x_1 + x_2 + x_3 = $ const. in Euclidian space $\mathbb{R}^3$

Treating each variable symmetrically, the barycentric coordinates $(x_1,\ x_2,\ x_3,\ x_4)$ are extended to 4-note chords as displayed (Figures 5a and 5b).

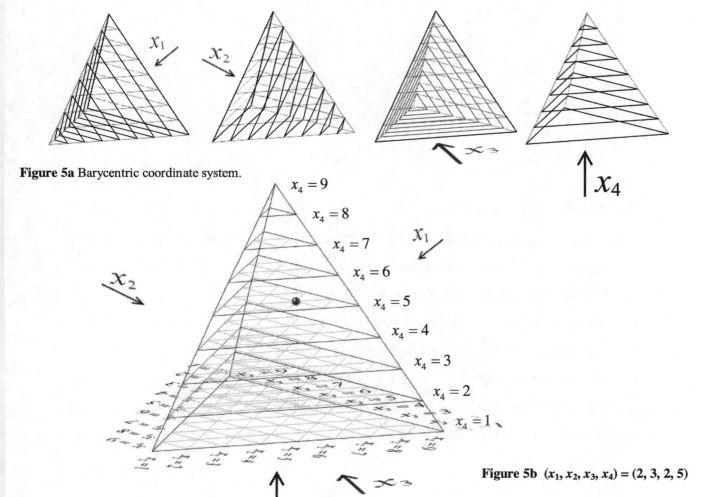

**Figure 5a** Barycentric coordinate system.

**Figure 5b** $(x_1, x_2, x_3, x_4) = (2, 3, 2, 5)$

5

The musical inversions of the same chord have the same intervallic distances in the clock notation, and this equivalence is represented in our interval notation by what we call the **orbit of a chord** defined in the following way. Given a chord $(x_1, x_2, x_3)$, a musical inversion of the same chord can be obtained by the cyclic permutation of the intervals; that is, the cyclic permutation $\sigma$ acting on $(x_1, x_2, x_3)$ transforms into the chord $(x_2, x_3, x_1)$. For example, the action of $\sigma$ on the chord $(4, 3, 5)$ produces chords $(3, 5, 4)$ and $(5, 4, 3)$. By connecting these points in the barycentric coordinate grid, we obtain an equilateral triangle which we call the orbit of the chord $(4, 3, 5)$ and we represent this by $\overline{(4, 3, 5)}$. The orbit of the triad is an equilateral consisting of vertices with the cyclic permutations of the chord intervals, and is centered at $(4, 4, 4)$ which is the barycenter of the coordinate system. Similarly $(3, 2, 7)$, $(2, 7, 3)$, and $(7, 3, 2)$ constitute the orbit $\overline{(3, 2, 7)}$. (See Figure 6.)

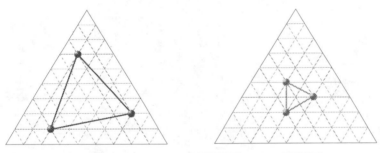

**Figure 6** Orbit of chords $\overline{(3, 2, 7)}$ and $\overline{(4, 3, 5)}$

## Metric and the Evenness Index

By introducing an appropriate metric in $\mathcal{T}^n$ we can inscribe orbital chords (a chord and its cyclic permutations) in concentric spheres centered at the barycenter $B$. The radii of these spheres will generate our evenness index under a very simple principle:

Chords inscribed in spheres of small radii will be close to the barycenter and thus, display a more even tonal distribution (consonance). At the same time chords approaching the outbound of the grid will display a higher content of clustering notes (dissonance).

In the case $n = 3$ we introduce a metric that takes the grid size to be of unit length along the side of a triangle, and the constituent chords located at the vertices to be equidistant from the barycenter. Given two triads $P$ and $Q$ with coordinates $\left(x_1, x_2, x_3\right)$ and $\left(y_1, y_2, y_3\right)$ respectively, we define:

$$\delta(P,Q) = \sqrt{\frac{\left(x_1 - y_1\right)^2 + \left(x_2 - y_2\right)^2 + \left(x_3 - y_3\right)^2}{2}}$$

If we define the evenness index $\|P\| = \delta(P, B)$, where $B = (4, 4, 4)$ is the barycenter, we can successfully measure how far a chord is from being an even intervallic partition of the octave. Higher $\|P\|$ values correspond to higher levels of dissonance, thus circumscribing circles of different $\delta$-radii centered at $B$ correspond to a proper classification of dissonance associated with the triads represented by equilateral triangles.

### Examples

The major and minor chords $(4, 3, 5)$ and $(3, 4, 5)$ have orbits that are inscribed in the smallest possible circle of $\delta$-radius equal to unity, making them the most consonant triads. (See Figure 7a.)

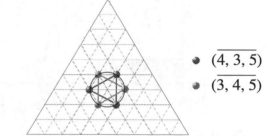

**Figure 7a** Orbits of chords $\overline{(4, 3, 5)}$ and $\overline{(3, 4, 5)}$.

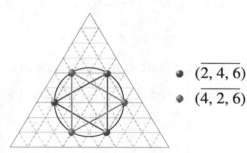

The orbit of chords $\overline{(2, 4, 6)}$ and $\overline{(4, 2, 6)}$ are both inscribed in a circle of $\delta$-radius 2. (See Figure 7b.)

**Figure 7b** Orbits of chords $\overline{(2, 4, 6)}$ and $\overline{(4, 2, 6)}$.

The cluster $\overline{(1, 1, 10)}$ is the most dissonant triad with a $\delta$-radius $3\sqrt{3}$. (See Figure 7c.)

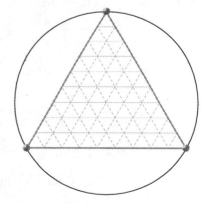

**Figure 7c** Orbit of chord $\overline{(1, 1, 10)}$.

7

The barycentric chord $B = (4, 4, 4)$ represents the augmented chord, and being the center of the coordinate system, it is not properly inscribed as an equilateral triangle in any $\delta$-circle. This is one of the several degenerated chords we encounter later in section 4. (See Figure 7d.) If we consider $B$ to be inscribed in a circle of $\delta$-radius zero, it would appear to be the most consonant triad which it is not, however its strategic position is essential for the proper development of this theory.

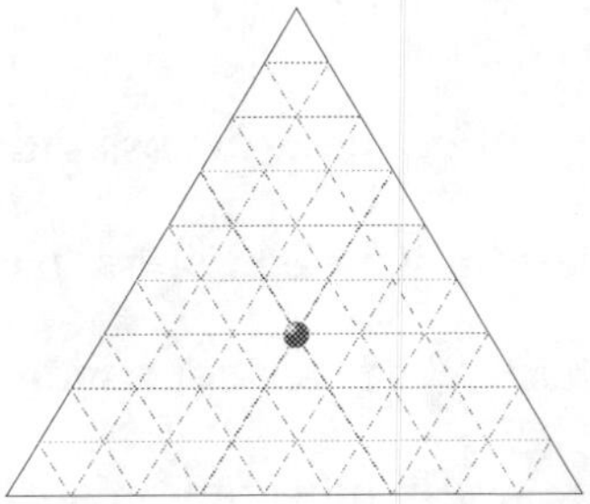

Figure 7d The barycentric chord $\overline{(4, 4, 4)}$.

## Properties of Chords

Now if we have a triad $\left(x_1, x_2, x_3\right)$ we consider its reflection $\left(x_3, x_2, x_1\right)$ which we call the transformation **inversion $I$**. Notice that $\left(x_1, x_2, x_3\right)$ and $I\left(x_1, x_2, x_3\right)$ are inscribed in the same $\delta$-radius circle and their orbits are symmetric with respect to any of the three bisectrices of the underlying triangular coordinate grid. Chords such as $\overline{(2, 5, 5)}$ are invariant under inversion since $I\overline{(2,5,5)} = \overline{(2,5,5)}$, and are called **$I$-symmetric.** Chords that are distinct and are related by inversion are called **co-$I$-symmetric.** $\overline{(2, 3, 7)}$ and $\overline{(3, 2, 7)}$ are co-$I$-symmetric and are displayed in Figure 8a.

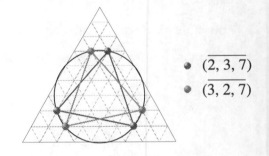

$\bullet$ $\overline{(2, 3, 7)}$
$\bullet$ $\overline{(3, 2, 7)}$

**Figure 8a**
co-$I$-symmetric chords $\overline{(2, 3, 7)}$ and $\overline{(3, 2, 7)}$.

Some chords like $\overline{(2, 5, 5)}$ and $\overline{(3, 3, 6)}$ are inscribed in the same $\delta$-circle while not related by inversion. We call these chords **$Y$-symmetric** and we treat them in great detail in section 2. (See Figure 8b.)

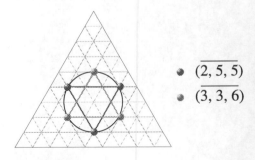

$\bullet$ $\overline{(2, 5, 5)}$
$\bullet$ $\overline{(3, 3, 6)}$

**Figure 8b**
$Y$-symmetric chords $\overline{(2, 5, 5)}$ and $\overline{(3, 3, 6)}$.

## Enrichments and Reductions

If we have a triad $(x_1, x_2, x_3)$ and we add a new note to the chord, this addition partitions one of the original intervals into two, creating a new 4-note chord of the form $(j, x_1 - j, x_2, x_3)$. This new chord is called an **enrichment** of the original $(x_1, x_2, x_3)$. Reciprocally if we start with a 4-note chord $(x_1, x_2, x_3, x_4)$ and we eliminate one of its constituent notes, the resulting triad will be the conjoining of two adjacent intervals in the original 4-note chord, for example $(x_1, x_2 + x_3, x_4)$. This new chord is referred as a **reduction** of the original chord. Enrichments and reductions are introduced in section 3. We study $I$-symmetric properties, and their evenness (consonance) implications.

## Degeneration

The 4-note case does not present a uniform scenario for the orbits of chords like the case with the triads. Most chords in the 4-note case are represented by isosceles tetrahedral orbits (see Figure 9). Some chords, due to their intrinsic symmetries collapse into two diametrically opposite points across the sphere of $\delta$-radius $\|P\|$ (called **polar chords**) and some chords collapse into a coplanar square (called **equatorial chords**). (See Figures 10a and 10b.) A thorough treatment of this phenomenon is covered in section 4. Here, linear algebraic methods, in particular techniques involving circulant matrices prove to be invaluable.

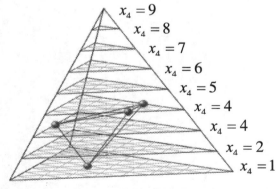

**Figure 9** Chord $\overline{(1, 4, 3, 4)}$

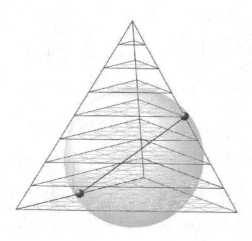 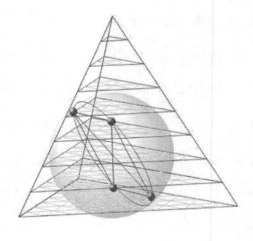

**Figure 10a** Polar chord $\overline{(1, 5, 1, 5)}$   **Figure 10b** Equatorial chord $\overline{(4, 1, 2, 5)}$

## Chord Complementation

Finally, in section 5 we treat the intervallic approach to chord complementation by introducing a very interesting idempotent map. It will be proven that two complementary chords are structurally closely related: The addition (enrichments) or elimination (reduction) of a few strategically chosen notes transform a chord to its complement.

# 1. Interval Space $\mathcal{T}^n$

Consider $n$ (where $n \geq 2$) independent points $\upsilon_1, \upsilon_2, \cdots, \upsilon_n$ in $\mathbb{R}^{n-1}$ (i.e. the vectors $\upsilon_2 - \upsilon_1, \upsilon_3 - \upsilon_1, \cdots, \upsilon_n - \upsilon_1$ are linearly independent in $\mathbb{R}^{n-1}$). Assume these points to be equidistant, and consider $x_1, x_2, \ldots, x_n$ discrete barycenter coordinates respect to $\upsilon_1, \upsilon_2, \cdots, \upsilon_n$.

(Namely, $0 < x_i < 12$, $x_i \in \mathbb{N}$ and $\sum_{i=1}^{n} x_i = 12$.) The set of points $P = \sum_{i=1}^{n} x_i \upsilon_i$ generated by these

coordinates determine a symmetric grid $\mathcal{T}^n$ of $\binom{12}{n} \dfrac{n}{12}$ points in $\mathbb{R}^{n-1}$. Any point in the grid is

uniquely represented by an ordered $n$-tuple $(x_1, x_2, \cdots, x_n)$ of its barycentric coordinates.
(See Figures 1.1a and 1.1b for cases $n = 3$, $n = 4$.)

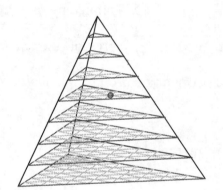

**Figure 1.1a** $(x_1, x_2, x_3) = (3, 2, 7)$ in $\mathcal{T}^3$    **Figure 1.1b** $(x_1, x_2, x_3, x_4) = (2, 3, 2, 5)$ in $\mathcal{T}^4$

We define a metric $\delta$ on $\mathcal{T}^n$ by:

$$\delta(P,Q) = \sqrt{\frac{\sum_{i=1}^{n}(x_i - y_i)^2}{2}}$$    where $P = (x_1, x_2, \ldots, x_n)$, $Q = (y_1, y_2, \ldots, y_n)$

## Examples

Given chords $P = (4, 3, 5)$ and $Q = (1, 3, 8)$ in $\mathcal{T}^3$,

$$\delta(P,Q) = \sqrt{\frac{(4-1)^2 + (3-3)^2 + (5-8)^2}{2}} = 3$$

Given chords $P = (2, 3, 1, 6)$ and $Q = (2, 1, 3, 7)$ in $\mathscr{T}^4$,

$$\delta(P,Q) = \sqrt{\frac{(2-2)^2 + (3-1)^2 + (1-2)^2 + (6-7)^2}{2}} = \sqrt{3}$$

The metric space $(\mathscr{T}^n, \delta)$ is called the **interval space** of $n$-chords.

Consider now $\Sigma^n$ a cyclic group of permutations generated by $\sigma$ acting on $\mathscr{T}^n$ as:

$$\sigma(x_1, x_2, \ldots, x_n) = (x_2, x_3, \ldots, x_n, x_1)$$

The quotient space emerged by this action is notated by $\mathscr{T}^n/\Sigma^n$.

The natural epimorphism:

$$p: \mathscr{T}^n \to \mathscr{T}^n/\Sigma^n \text{ defined by } p(x_1, x_2, \ldots, x_n) = \overline{(x_1, x_2, \ldots, x_n)}$$

where $\overline{(x_1, x_2, \ldots, x_n)}$ represents the equivalence class in $\mathscr{T}^n/\Sigma^n$ integrated by the elements of the cyclic permutation group $\Sigma^n$ acting on $(x_1, x_2, \ldots, x_n)$. This class

$$\overline{(x_1, x_2, \ldots, x_n)} = \left\{ (x_1, x_2, \ldots, x_n), \sigma(x_1, x_2, \ldots, x_n), \sigma^2(x_1, \ldots, x_n), \ldots, \sigma^{n-1}(x_1, \ldots, x_n) \right\}$$

is called the **orbit** of $(x_1, x_2, \ldots, x_n)$.

**Examples**

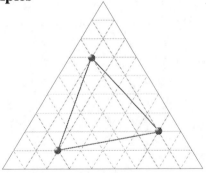

Figure 1.2a $\overline{(3,2,7)}$ in $\mathscr{T}^3$

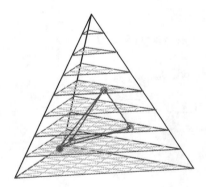

Figure 1.2b $\overline{(2,3,2,5)}$ in $\mathscr{T}^4$

Notice that if $(x_1, x_2, \ldots, x_n)$ and $(y_1, y_2, \ldots, y_n)$ are in the same class of $\mathscr{T}^n/\Sigma^n$ then

$$\overline{(x_1, x_2, \ldots, x_n)} = \overline{(y_1, y_2, \ldots, y_n)}.$$

We extend the metric $\delta$ to $\mathcal{T}^n/\Sigma^n$ in the following manner:

$$\delta^\Sigma\left(\overline{(x_1, x_2, \ldots, x_n)}, \overline{(y_1, y_2, \ldots, y_n)}\right) = \min_{\tau, \mu \in \Sigma^n} \delta\left((\tau(x_1, x_2, \ldots, x_n), \mu(y_1, y_2, \ldots, y_n))\right)$$

$$= \min_{\mu \in \Sigma^n} \delta\left((x_1, x_2, \ldots, x_n), \mu(y_1, y_2, \ldots, y_n)\right)$$

The new metric space $\left(\mathcal{T}^n/\Sigma^n, \delta^\Sigma\right)$ is called the orbital space of $n$-chords.


**Example**

The distance between the orbit of the major triad $\overline{(4, 3, 5)}$ and that of the minor triad $\overline{(3, 4, 5)}$

is

$$\delta^\Sigma\left(\overline{(4, 3, 5)}, \overline{(3, 4, 5)}\right) = \min\left(\delta\left((4, 3, 5), (3, 4, 5)\right), \delta\left((4, 3, 5), (4, 5, 3)\right), \delta\left((4, 3, 5), (5, 3, 4)\right)\right)$$

$$= \min(1, 2, 1) = 1$$


Let's introduce a new operator $I$, the inversion operator acting on $\mathcal{T}^n$ as:

$$I(x_1, x_2, \ldots, x_n) = (x_n, x_{n-1}, \ldots, x_1)$$

Notice $I^2 = e$, the identity. Example: If $P = (2, 3, 1, 6)$, then $I(P) = (6, 1, 3, 2)$.


The group $\Phi^n$ generated by $\sigma$ and $I$ is a non-abelian group of $2n$ elements.


**Example**

Consider the group generated by $\sigma$ and $I$ and form $\Phi^3$, which is a six element non-abelian group. Notice:

$$I \circ \sigma = \sigma^2 \circ I$$

and $\quad I \circ \sigma^2 = \sigma \circ I$

The group table for $\Phi^3$ is displayed below:

| $e$ | $\sigma$ | $\sigma^2$ | $I$ | $I \circ \sigma$ | $I \circ \sigma^2$ |
|---|---|---|---|---|---|
| $\sigma$ | $\sigma^2$ | $e$ | $I \circ \sigma^2$ | $I$ | $I \circ \sigma$ |
| $\sigma^2$ | $e$ | $\sigma$ | $I \circ \sigma$ | $I \circ \sigma^2$ | $I \circ \sigma$ |
| $I$ | $I \circ \sigma$ | $I \circ \sigma^2$ | $e$ | $\sigma$ | $\sigma^2$ |
| $I \circ \sigma$ | $I \circ \sigma^2$ | $I$ | $\sigma^2$ | $e$ | $\sigma$ |
| $I \circ \sigma^2$ | $I$ | $I \circ \sigma$ | $\sigma$ | $\sigma^2$ | $e$ |

## Example

Consider the group generated by $\sigma$ and $I$ and form $\Phi^4$, which is an 8 element non-abelian group.

Notice:

$$\sigma \circ I = I \circ \sigma^3$$

$$I \circ \sigma = \sigma^3 \circ I$$

$$I \circ \sigma^2 = \sigma^2 \circ I, \text{ and the group table for } \Phi^4 \text{ is}$$

| $e$ | $\sigma$ | $\sigma^2$ | $\sigma^3$ | $I$ | $I \circ \sigma$ | $I \circ \sigma^2$ | $I \circ \sigma^3$ |
|---|---|---|---|---|---|---|---|
| $\sigma$ | $\sigma^2$ | $\sigma^3$ | $e$ | $I \circ \sigma^3$ | $I$ | $I \circ \sigma$ | $I \circ \sigma^2$ |
| $\sigma^2$ | $\sigma^3$ | $e$ | $\sigma$ | $I \circ \sigma^2$ | $I \circ \sigma^3$ | $I$ | $I \circ \sigma$ |
| $\sigma^3$ | $e$ | $\sigma$ | $\sigma^2$ | $I \circ \sigma$ | $I \circ \sigma^2$ | $I \circ \sigma^3$ | $I$ |
| $I$ | $I \circ \sigma$ | $I \circ \sigma^2$ | $I \circ \sigma$ | $e$ | $\sigma$ | $\sigma^2$ | $\sigma^3$ |
| $I \circ \sigma$ | $I \circ \sigma^2$ | $I \circ \sigma^3$ | $I$ | $\sigma^3$ | $e$ | $\sigma$ | $\sigma^2$ |
| $I \circ \sigma^2$ | $I \circ \sigma^3$ | $I$ | $I \circ \sigma$ | $\sigma^2$ | $\sigma^3$ | $e$ | $\sigma$ |
| $I \circ \sigma^3$ | $I$ | $I \circ \sigma$ | $I \circ \sigma^2$ | $\sigma$ | $\sigma^2$ | $\sigma^3$ | $e$ |

In general, for $k < n$

$$\sigma^k \circ I \, (x_1, x_2, \ldots, x_n) \;=\; \sigma^k (x_n, x_{n-1}, \ldots, x_1) \;=\; (x_{n-k}, x_{n-k-1}, \ldots, x_1, x_n, \ldots, x_{n-k+1})$$

$$I \circ \sigma^{n-k} (x_1, x_2, \ldots, x_n) \;=\; I \, (x_{n-k+1}, x_{n-k+2}, \ldots, x_n, x_1, \ldots, x_{n-k}) \;=\; (x_{n-k}, x_{n-k-1}, \ldots, x_1, x_n, \ldots, x_{n-k+1})$$

So for every chord $P$ in $\mathcal{T}^n$,

$$\sigma^k \circ I \, (P) \;=\; I \circ \sigma^{n-k} (P)$$

**Example**

If $P = (1, 2, 1, 2, 2, 1, 3)$

then

$$\sigma^5 \circ I(P) = \sigma^5 (3, 1, 2, 2, 1, 2, 1) = (2, 1, 3, 1, 2, 2, 1)$$

and

$$I \circ \sigma^2 (P) = I(1, 2, 2, 1, 3, 1, 2) = (2, 1, 3, 1, 2, 2, 1)$$

The above identity allows us to extend the inversion operator to $\mathcal{T}^n / \Sigma^n$ in the following way

$$I(\overline{P}) = \overline{I(P)} \quad \text{for every } P \text{ in } \mathcal{T}^n.$$

Consider a new quotient set $\mathcal{T}^n / \Phi^n$ and the natural epimorphism:

$$p \colon \mathcal{T}^n \to \mathcal{T}^n / \Phi^n$$

defined by

$$p(\, x_1, x_2, \ldots, x_n) = \overline{\overline{(x_1, x_2, \ldots, x_n)}}$$

where $\overline{\overline{(x_1, x_2, \ldots, x_n)}}$ represents the class in $\mathcal{T}^n / \Phi^n$ generated by the $2n$ elements of $\Phi^n$ acting on $(\, x_1, x_2, \ldots, x_n)$ with both cyclic permutation and inversions.

**Examples**

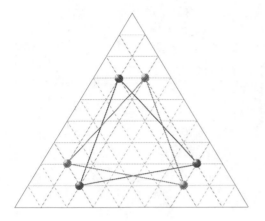

- $(3, 2, 7)$
- $I(3, 2, 7) = (2, 3, 7)$

**Figure 1.3a** $\overline{\overline{(3,2,7)}}$ in $\mathscr{T}^3/\Phi^3$

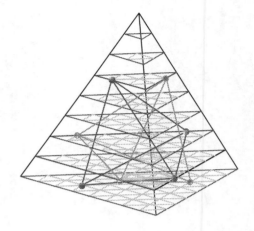

- $(3, 2, 1, 6)$
- $I(3, 2, 1, 6) = (1, 2, 3, 6)$

**Figure 1.3b** $\overline{\overline{(3,2,1,6)}}$ in $\mathscr{T}^4/\Phi^4$

Notice that $\overline{\overline{I(P)}} = \overline{\overline{P}}$ for any chord $P$ in $\mathscr{T}^n$.

In an analogous way we define a metric $\delta^\Phi$ on $\mathscr{T}^n/\Phi^n$

$$\delta^\Phi\left(\overline{\overline{(x_1, x_2, \ldots, x_n)}}, \overline{\overline{(y_1, y_2, \ldots, y_n)}}\right) = \min_{\tau, \mu \in \Phi^n} \delta\big(\tau(x_1, x_2, \ldots, x_n),\ \mu(y_1, y_2, \ldots, y_n)\big)$$

$$= \min_{\mu \in \Phi^n} \delta\big((x_1, x_2, \ldots, x_n),\ \mu(y_1, y_2, \ldots, y_n)\big)$$

**Example**

If $P = (4, 3, 5)$ and $Q = (3, 2, 7)$, then

$$\delta^\Phi\left(\overline{\overline{P}}, \overline{\overline{Q}}\right) = \min\begin{pmatrix} \delta\big((4, 3, 5),\ (3, 2, 7)\big),\ \delta\big((4, 3, 5),\ (7, 2, 3)\big),\ \delta\big((4, 3, 5),\ (2, 7, 3)\big), \\ \delta\big((4, 3, 5),\ (3, 7, 2)\big),\ \delta\big((4, 3, 5),\ (7, 3, 2)\big),\ \delta\big((4, 3, 5),\ (2, 3, 7)\big) \end{pmatrix}$$

$$= \min\left(\sqrt{3},\ \sqrt{7},\ 2\sqrt{3},\ \sqrt{13},\ 3,\ 2\right) = \sqrt{3}$$

The resulting metric space $\left(\mathcal{T}^n/\Phi^n, \delta^\Phi\right)$ is called the Forte space of $n$-chords. (Honoring the pioneering work of the music set theorist Allen Forte.)

**Evenness Index and Index of Dissonance**

As we mentioned in the introduction the highest consonance in musical chords occurs when the notes in the chord partition the octave in equal or almost equal intervals. Consequently, chords that contain notes that are clustered close together are considered very dissonant. Translating this principle to our musical space $(\mathcal{T}^n, \delta)$, chords that are close to the barycenter in the $\delta$-metric are more consonant, and reciprocally, chords that lie in the outerbounds of the grid are considerably more dissonant. This canonical representation of chords in the $\delta$-metric space allows for analytically quantifying the interval relationships of chords. It is then natural to regard this metric as a powerful and objective tool to measure dissonance in musical chords.

Given a chord $P = (x_1, x_2, ..., x_n)$ in $(\mathcal{T}^n, \delta)$ we define its evenness index or simply index

as: $\qquad \text{Index}(P) = \|P\| = \sqrt{\dfrac{\sum\limits_{i=1}^{n}(x_i - b)}{2}}$ where $b = \dfrac{12}{n}$ and the barycenter $B^n = (\overbrace{\dfrac{12}{n}, \dfrac{12}{n}, ..., \dfrac{12}{n}}^{n})$ is

a member of $\mathcal{T}^n$, if and only if, $n$ divides 12. Notice that for every $\sigma$ in $\Sigma^n$, $\|\sigma(P)\| = \|P\|$, and also $\|I(P)\| = \|P\|$, which allow us to extend the concept of index to the quotient spaces $(\mathcal{T}^n/\Sigma^n, \delta^\Sigma)$ and $(\mathcal{T}^n/\Phi^n, \delta^\Phi)$.

Since higher evenness indices correspond to more unevenness in the chord, it would be more appropriate perhaps, to label our index as the unevenness index instead of the evenness index. However we opted for the latter one, since we consider the unevenness term rather awkward.

In view of the fact that the maximum level of consonance does not occur at the barycenter itself, but instead occurs at the smallest $\delta$-circle centered at $B^n$ in $\mathcal{T}^n$, we propose two alternative definitions of dissonance/consonance.

Index of dissonance of $P$:

$$Index_d(P) = \left| \|P\| - \delta_{\min}^n \right|$$

where $\delta_{min}^{n}$ represents the radius (different from zero) of the smallest circle in the discrete coordinate system $(\mathcal{T}^{n}, \delta)$ (i.e. $\delta_{min}^{3}=1$, $\delta_{min}^{4}=1$, $\delta_{min}^{5} = \sqrt{\frac{3}{5}}$). Notice that in this expression the major and minor triad has a zero index of dissonance value and the barycenter (4,4,4) has a value of 1.

Still alternatively, we can consider an index of consonance as:

$$Index_{c}(P) = \alpha_{n} - Index_{d}(P)$$

where $\alpha_{n} = \delta_{max}^{n} - \delta_{min}^{n}$, is the difference between the maximum and minimum radii of concentric evenness circles in the space $(\mathcal{T}^{n}, \delta)$.

The result for the index of consonance is graphed as follows:

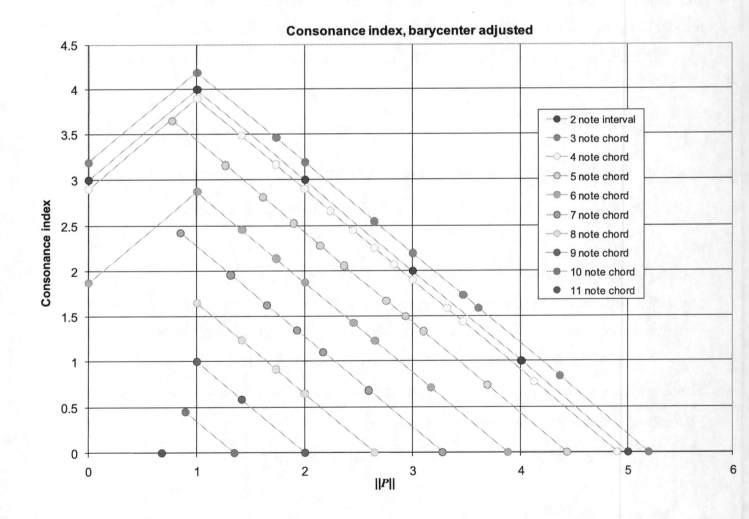

## Orbits and meshes

We now define the $k^{\text{th}}$ orbit of $P$ in $\mathcal{T}^n$ as:

$$\mathcal{O}_k(P) = \left\{ P,\ \sigma^k(P),\ \sigma^{2k}(P),\ \ldots,\ \sigma^{(n-1)k}(P) \right\}$$

Notice that $\mathcal{O}_k(P)$ has at most 12 elements and that $\mathcal{O}_1(P) = \overline{P}$. Since the sets $\left\{(x_i - x_{k+i}), i = 1, \ldots, n\right\}$ and $\left\{(x_{k+i} - x_{2k+i}), i = 1, \ldots, n\right\}$ are the same:

$$\delta(P, \sigma^k(P)) \;=\; \delta(\sigma^k(P), \sigma^{2k}(P)) \;=\; \delta(\sigma^{ik}(P), \sigma^{(i+1)k}(P))\ ,\, i = 2, \ldots,\, n-2$$

And this invariant quantity is called the mesh of $\mathcal{O}_k(P)$ and is notated $\widehat{\sigma_k}(P)$.

It is easy to check that $\widehat{\sigma_k}(P)\,(\tau(P)) = \widehat{\sigma_k}(P)$ for all $\tau$ in $\Sigma^n$. Also $\widehat{\sigma_k}(I(P)) = \widehat{\sigma_k}(P)$. And, as usual, we can extend these quantities to $\mathcal{T}^n/\Sigma^n$ and $\mathcal{T}^n/\Phi^n$.

The meshes $\widehat{\sigma_k}$ play an essential role in the visual geometric representation of chords in $\mathcal{T}^n/\Sigma^n$. For instance, in $\mathcal{T}^3/\Sigma^3$, every orbital chord $\overline{P}$ is represented by an equilateral triangle with vertices at $P,\ \sigma(P),\ \sigma^2(P)$, and with sides of length $\widehat{\sigma_1}(\overline{P})$.

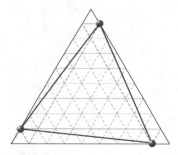

**Figure 1.4**

$P = \overline{(1,\, 2,\, 9)}$ with $\widehat{\sigma_1}(\overline{P}) = \sqrt{57}$

In $\mathscr{T}^4/\Sigma^4$ every orbital chord $\overline{P}$ is represented by an isosceles tetrahedron (there are a few interesting exceptions) with vertices at $P$, $\sigma(P)$, $\sigma^2(P)$, $\sigma^3(P)$ and whose faces are congruent isosceles triangles with edges $\widehat{\sigma_1}(\overline{P})$, $\widehat{\sigma_1}(\overline{P})$, $\widehat{\sigma_2}(\overline{P})$.

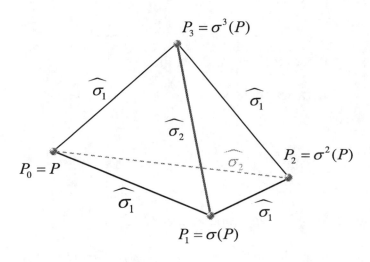

$P_3 = \sigma^3(P)$

$\widehat{\sigma_1}$

$\widehat{\sigma_1}$

$\widehat{\sigma_2}$

$\widehat{\sigma_2}$

$P_0 = P$

$P_2 = \sigma^2(P)$

$\widehat{\sigma_1}$

$\widehat{\sigma_1}$

$P_1 = \sigma(P)$

**Figure 1.5**
Orbital of chord $P$ represented as a 3-simplex (tetrahedron), with vertices at $P_k = \sigma^k(P)$.

**Example**

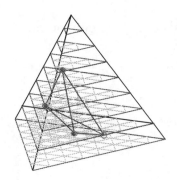

**Figure 1.6**
In $\mathscr{T}^4$, $P = \overline{(1, 2, 4, 5)}$ is displayed with $\widehat{\sigma_1}(\overline{P}) = \sqrt{11}$, $\widehat{\sigma_2}(\overline{P}) = 3\sqrt{2}$

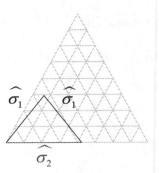

$\widehat{\sigma_1}$ $\widehat{\sigma_1}$

$\widehat{\sigma_2}$

**Figure 1.7**
Each face of the tetrahedron for $P = \overline{(1, 2, 4, 5)}$ is an isosceles triangle with sides of lengths $\widehat{\sigma_1}(\overline{P}) = \sqrt{11}$, $\widehat{\sigma_2}(\overline{P}) = 3\sqrt{2}$

The higher dimensions are harder to visualize. However, the $\widehat{\sigma_k}$ gives us a clear insight of the geometrical properties of the chord. The natural generalization of the tetrahedron to higher dimensions would be the topological concept of a simplex.

In $\mathcal{T}^n/\Sigma^n$ a chord $P$ is represented by an $n-1$ simplex in $\mathbb{R}^{n-1}$ with vertices at $P$, $\sigma(P)$, $\sigma^2(P)$, ..., $\sigma^{n-1}(P)$. The 2-skeleton of this simplex is the simplicial 2 dimensional complex formed by $\binom{n}{3}$ triangles. The different triangles in the skeleton are called the generating triangles. These triangles are determined by $\widehat{\sigma_1}$, $\widehat{\sigma_2}$,..., $\widehat{\sigma_k}$, where $k = \left[\!\left[\dfrac{n}{2}\right]\!\right]$, since $\widehat{\sigma_i} = \widehat{\sigma_j}$ if $i+j=n$.

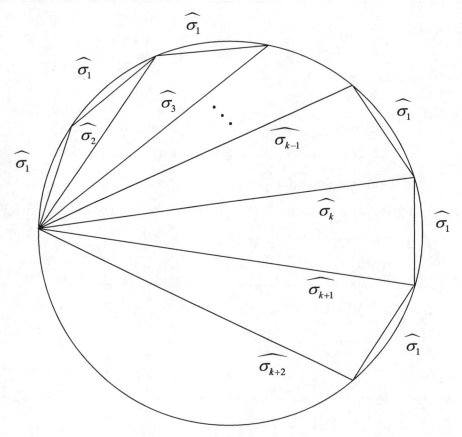

Notice that if $k \neq \dfrac{n}{2}$ we have two adjacent $\widehat{\sigma_k}$.

**Figure 1.8**
Planar representation of the $n$-chord

21

A chord in $\mathcal{T}^n$ produces:

| | | | |
|---|---|---|---|
| $n$ triangles | $\widehat{\sigma_i}\ \widehat{\sigma_i}\ \widehat{\sigma_j}$ | if $2i+j=n$ | $,i\neq j$ |
| $n$ triangles | $\widehat{\sigma_i}\ \widehat{\sigma_i}\ \widehat{\sigma_j}$ | if $2i+j<n$ | and $j=2i$ |
| $2n$ triangles | $\widehat{\sigma_i}\ \widehat{\sigma_j}\ \widehat{\sigma_k}$ | if $i+j+k=n$ | $,i\neq j\neq k$ |
| $2n$ triangles | $\widehat{\sigma_i}\ \widehat{\sigma_j}\ \widehat{\sigma_k}$ | if $i+j+k<n$ | and $k-j=i$ |
| $n/3$ triangles | $\widehat{\sigma_{n/3}}\ \widehat{\sigma_{n/3}}\ \widehat{\sigma_{n/3}}$ | if $n\,|\,3$ | |

For instance a hexachord has 3 generating triangles: $\widehat{\sigma_1}\ \widehat{\sigma_1}\ \widehat{\sigma_2}$, $\widehat{\sigma_1}\ \widehat{\sigma_2}\ \widehat{\sigma_3}$, and $\widehat{\sigma_2}\ \widehat{\sigma_2}\ \widehat{\sigma_2}$.

The 2-skeleton of the hexachord has:

| | | |
|---|---|---|
| 6 triangles | $\widehat{\sigma_1}\ \widehat{\sigma_1}\ \widehat{\sigma_2}$ | |
| 12 triangles | $\widehat{\sigma_1}\ \widehat{\sigma_2}\ \widehat{\sigma_3}$ | |
| 2 triangles | $\widehat{\sigma_2}\ \widehat{\sigma_2}\ \widehat{\sigma_2}$ | |

**Examples**

### Skeleton Triangles

| $n$-chord | Number of generating triangles | Total number of triangles |
|---|---|---|
| 3 | 1 | 1 |
| 4 | 1 | 4 |
| 5 | 2 | 10 |
| 6 | 3 | 20 |
| 7 | 4 | 35 |
| 8 | 5 | 56 |
| 9 | 7 | 84 |
| 10 | 8 | 120 |
| 11 | 10 | 165 |
| 12 | 12 | 220 |

## 2. *I*-symmetry, co-*I*-symmetry, and *Y*-symmetry

Two Chords $P$ and $Q$ in $\mathcal{T}^n$ are called co-*I*-symmetric if and only if $\overline{P} = I(\overline{Q})$. A chord $P$ in $\mathcal{T}^n$ is called *I*-symmetric if it is co-*I*-symmetric with itself, i.e. $\overline{P} = I(\overline{P})$. If two chords $P$ and $Q$ are co-*I*-symmetric there exists at least one $\overline{k}$ $n$ such that

$$\sigma^k(P) = I(Q)$$

The minimum of these $k$'s is called the inversion exponent of $P$ with respect to $Q$.

### Examples
Given $P = (1, 2, 4, 5)$ and $Q = (4, 2, 1, 5)$ in $\mathcal{T}^4$, $P$ and $Q$ and co-*I*-symmetric and the inversion exponent is $P$ with respect to $Q$ is 3, since

$$\sigma^3(P) = (5, 1, 2, 4) = I(Q)$$

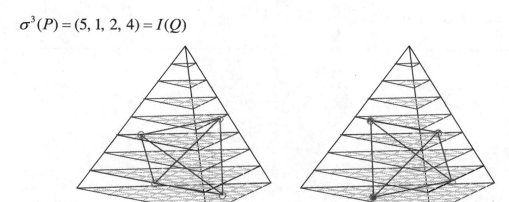

**Figure 2.1** $P = (1, 2, 4, 5)$ and $Q = (4, 2, 1, 5)$ in $\mathcal{T}^4$.

Given $P = (1, 2, 3, 1, 2, 3)$ and $Q = (2, 1, 3, 2, 1, 3)$ in $\mathcal{T}^6$, $P$ and $Q$ and co-*I*-symmetric and we have exponents 2 and 5 verifying

$$\sigma^2(P) = \sigma^5(P) = (3, 1, 2, 3, 1, 2) = I(Q)$$

Thus 2 is the inversion exponent of $P$ with respect to $Q$.

Define the *I*-symmetry index of a chord $P$ inscribed in the circle of $\delta$-radius $\|P\|$:

$$i\left(\overline{P}\right) = \frac{\delta^\Sigma\left(\overline{P}, I\left(\overline{P}\right)\right)}{\|P\|}$$

Notice that $i(\overline{P}) = i(I(\overline{P}))$. *I*-symmetric chords have *I*-symmetry index of zero. Chords with higher *I*-symmetry index show a more distinctive sonority quality between them and their inversions.

Consider the following: Given $P = (x_1, \cdots, x_n)$ and $Q = (y_1, \cdots, y_n)$ in $\mathcal{T}^n$ and $\overline{P} = I(\overline{Q})$. Assume without loss of generality that the inversion exponent $k = 0$, then we have

$$x_1 = y_n, \quad x_2 = y_{n-1}, \quad \cdots, \quad x_n = y_1 \qquad \text{or} \qquad x_i = y_{n-i+1}$$

Therefore

$$2\|P\|^2 = \sum_{i=1}^n (x_i - b)^2 = \sum_{i=1}^n (y_{n-i+1} - b)^2 = \sum_{j=1}^n (y_j - b)^2 = 2\|Q\|^2$$

where $b$ is the barycentric coordinate $\dfrac{12}{n}$. Hence $\|P\| = \|Q\|$.

Similarly,

$$2\widehat{\sigma_k}^2(P) = \sum_{i=1}^n (x_i - x_{k+i})^2 = \sum_{i=1}^n (y_{n-i+1} - y_{n-k-i+1})^2 = \sum_{j=1}^n (y_j - y_{j+k})^2 = 2\widehat{\sigma_k}^2(Q)$$

since

$$(n-i+1) - (n-k-i+1) = k$$

and

$$\widehat{\sigma_k}(P) = \widehat{\sigma_k}(Q) \qquad \text{for } 1 \le k \le n$$

and we have proved that if two chords are co-*I*-symmetric, their indexes and orbital meshes are the same.

We are going to introduce a new concept, *Y*-symmetry, that originated in the study of chords that share the same index and the same orbital meshes but they are not *I*-symmetry related.
Consider a map $Y : \mathcal{T}^n \to \mathbb{Z}^n$ defined by

$$Y(x_1, \cdots, x_n) = (x_2 - x_1, x_3 - x_2, \cdots, x_n - x_{n-1}, x_1 - x_n)$$

Two chords $P$ and $Q$ in $\mathcal{T}^n$ are said to be *Y*-symmetric if and only if there exists a chord $Q'$ in $\overline{\overline{Q}}$ such that $Y(P) = -Y(Q')$.

**Examples**

$P = (3, 2, 2, 5)$ and $Q = (3, 1, 4, 4)$ are *Y*-symmetric since $Q' = (3, 4, 4, 1) \in \overline{\overline{Q}}$ and

$$Y(P) = (-1, 0, 3, -2) = -Y(Q')$$

24

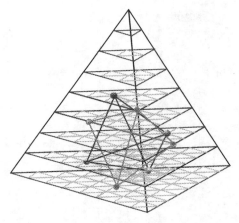

**Figure 2.2** $P = (3, 2, 2, 5)$, and $Q = (3, 1, 4, 4)$ in $\mathcal{T}^4$

$P = (2, 5, 5)$ and $Q = (3, 3, 6)$ are $Y$-symmetric since $Q' = (6, 3, 3) \in \overline{\overline{Q}}$ and

$$Y(P) = (3, 0, -3) = -Y(Q')$$

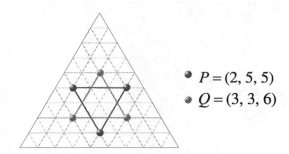

$P = (2, 5, 5)$

$Q = (3, 3, 6)$

**Figure 2.3** $P = (2, 5, 5)$, and $Q = (3, 3, 6)$ in $\mathcal{T}^3$

Notice that if $P$ and $Q$ are co-$I$-symmetric then they are $Y$-symmetric. Effectively we arrive at the following. Assume $P$ and $Q$ as above with the inversion exponent $k = 0$. From the relation $x_i = y_{n-i+1}$ we obtain:

$$x_{i+1} - x_i = y_{n-(i+1)+1} - y_{n-i+1} = y_{n-i} - y_{n-i+1} = -(y_{n-i+1} - y_{n-i})$$

Hence, $Y(P) = -Y(Q')$, where $Q' = I(Q)$, and $P$ and $Q$ are $Y$-symmetric.

Even if $Y$-symmetry is weaker than $I$-symmetry, the condition is very strong. We are going to prove that if two chords $P$ and $Q$ in $\mathcal{T}^n$ are $Y$-symmetric then they share the same orbital meshes and the same index.

Assume without any loss of generality that $P = (x_1, \cdots, x_n)$ and $Q' = Q = (y_1, \cdots, y_n)$ in $\mathcal{T}^n$ and $Y(P) = -Y(Q)$, that is,

$$(x_2 - x_1, x_3 - x_2, \cdots, x_n - x_{n-1}, x_1 - x_n) = -(y_2 - y_1, y_3 - y_2, \cdots, y_n - y_{n-1}, y_1 - y_n)$$

which gives us:

$$\begin{cases} x_2 - x_1 & = & y_1 - y_2 \\ x_3 - x_2 & = & y_2 - y_3 \\ x_4 - x_3 & = & y_3 - y_4 \\ & \vdots & \\ x_k - x_{k-1} & = & y_{k-1} - y_k \\ x_{k+1} - x_k & = & y_k - y_{k+1} \\ & \vdots & \\ x_n - x_{n-1} & = & y_{n-1} - y_n \\ x_1 - x_n & = & y_n - y_1 \end{cases} \qquad (2.1)$$

From (2.1) it follows that $\widehat{\sigma_1}(P) = \widehat{\sigma_1}(Q)$. Adding two consecutive rows in (2.1) gives:

$$\begin{cases} x_3 - x_1 & = & y_1 - y_3 \\ x_3 - x_2 & = & y_2 - y_3 \\ & \vdots & \\ x_{k+2} - x_k & = & y_k - y_{k+2} \\ & \vdots & \\ x_n - x_{n-1} & = & y_{n-1} - y_n \\ x_2 - x_n & = & y_n - y_2 \end{cases}$$

and thus $\widehat{\sigma_2}(P) = \widehat{\sigma_2}(Q)$.

Analogously adding $k$ consecutive rows, we get

$$\begin{cases} x_{k+1} - x_1 & = & y_1 - y_{k+1} \\ x_{k+2} - x_2 & = & y_2 - y_{k+2} \\ & \vdots & \end{cases}$$

and thus $\widehat{\sigma_k}(P) = \widehat{\sigma_k}(Q)$.

Let's prove now the equality of the indices.
Since

$$\left(\sum_{i=1}^{n} x_i\right) \cdot \left(\sum_{i=1}^{n} x_i\right) = \left(\sum_{i=1}^{n} y_i\right) \cdot \left(\sum_{i=1}^{n} y_i\right) = 12^2 = 144$$

Expanding the left hand term we obtain

$$\sum_{i=1}^{n} x_i^2 + 2\sum_{i=1}^{n} x_i x_{i+1} + 2\sum_{i=1}^{n} x_i x_{i+2} + \cdots + ((2))\sum_{i=1}^{n} x_i x_{i+k} \quad , \text{where } k = \left\lfloor \frac{n}{2} \right\rfloor$$

and

$$((2)) = \begin{cases} 2 & \text{if } n \text{ is odd} \\ 1 & \text{if } n \text{ is even} \end{cases}$$

Now,

$$2\widehat{\sigma_1}^2(P) = \sum_{i=1}^{n} (x_i - x_{i+1})^2 = 2\sum_{i=1}^{n} x_i^2 - 2\sum_{i=1}^{n} x_i x_{i+1} \quad \Rightarrow \quad 2\sum_{i=1}^{n} x_i x_{i+1} = 2\sum_{i=1}^{n} x_i^2 - 2\widehat{\sigma_1}^2(P)$$

And for $j = 1, 2, \cdots, k$ we get

$$2\sum_{i=1}^{n} x_i x_{i+j} = 2\sum_{i=1}^{n} x_i^2 - 2\widehat{\sigma_j}^2(P)$$

Similarly,

$$2\sum_{i=1}^{n} y_i y_{i+j} = 2\sum_{i=1}^{n} y_i^2 - 2\widehat{\sigma_j}^2(Q)$$

By back substituting to above, and equating left hand and right hand sides we have

$$\sum_{i=1}^{n} x_i^2 + 2k\sum_{i=1}^{n} x_i^2 - 2\sum_{i=1}^{k} \widehat{\sigma_j}^2(P) = \sum_{i=1}^{n} y_i^2 + 2k\sum_{i=1}^{n} y_i^2 - 2\sum_{i=1}^{k} \widehat{\sigma_j}^2(Q)$$

Since $\widehat{\sigma_i}(P) = \widehat{\sigma_i}(Q)$ for all $i = 1, \cdots, k$ we obtain

$$\boxed{\sum_{i=1}^{n} x_i^2 = \sum_{i=1}^{n} y_i^2} \quad (2.2) \quad \text{and consequently} \quad \boxed{\sum_{i=1}^{n} x_i x_{i+j} = \sum_{i=1}^{n} y_i y_{i+j} \ , \ j = 1, \cdots, k} \quad (2.3)$$

Let now $b = \dfrac{12}{n}$ be the barycentric coordinate in $\mathcal{T}^n$. Then,

$$2\widehat{\sigma_1}^2(P) = \sum_{i=1}^{n}(x_i - x_{i+1})^2 = \sum_{i=1}^{n}\left((x_i - b) - (x_{i+1} - b)\right)^2 = 4\|P\|^2 - 2\sum_{i=1}^{n}x_i x_{i+1} + 48b - 2nb^2$$

and $\quad 2\widehat{\sigma_1}^2(Q) = 4\|Q\|^2 - 2\sum_{i=1}^{n}x_i x_{i+1} + 48b - 2nb^2$

and by (2.3) we obtain: $\|P\| = \|Q\|$.

It is important to remark that the above proof enables us to state the important result:

> If two chords $P$ and $Q$ in $\mathcal{T}^n$ share the same orbital meshes $\widehat{\sigma}_k$ then their indices must be equal.

There are some chords in dimensions $n \geq 5$, that share the same indices and orbital meshes but they are not $Y$-symmetric. These chords, however, satisfy the weaker conditions (2.2) or (2.3) above.

**Example**

Chords $(1, 2, 1, 4, 4)$ and $(2, 2, 1, 2, 5)$ in $\mathcal{T}^5$ have index $\sqrt{\dfrac{23}{5}}$, and they both have $\widehat{\sigma}_1 = \sqrt{10}$ and $\widehat{\sigma}_2 = \sqrt{13}$.

Since $\quad Y(1, 2, 1, 4, 4) = (1, -1, 3, 0, -3)$, $\quad$ and $\quad Y(2, 2, 1, 2, 5) = (0, -1, 1, 3, -3)$ the chords are not $Y$-symmetric, but they satisfy both conditions (2.2) and (2.3).

$$1^2 + 2^2 + 1^2 + 4^2 + 4^2 = 2^2 + 2^2 + 1^2 + 2^2 + 5^2 = 38$$

$$1(2) + 2(1) + 1(4) + 4(4) + 4(1) = 2(2) + 2(1) + 1(2) + 2(5) + 5(2) = 28 \qquad (j = 1)$$

$$1(1) + 2(4) + 1(4) + 4(1) + 4(2) = 2(1) + 2(2) + 1(5) + 2(2) + 5(2) = 25 \qquad (j = 2)$$

# 3. Enrichments and Reductions

Given $(x_1, x_2, \cdots, x_i, \cdots, x_n) \in \mathcal{T}^n$, the interval $x_i$ can be partitioned into two intervals through the addition of one more note in the chord.

The interval $x_i$ is separated into two intervals $j$ and $x_i - j$, where $1 \leq j \leq x_i - 1$. An **enrichment** $f_i^j$ is a map from $\mathcal{T}^n$ to $\mathcal{T}^{n+1}$ defined by:

$$f_i^j(x_1, x_2, \cdots, x_i, \cdots, x_n) = (x_1, x_2, \cdots, x_{i-1}, j, x_i - j, x_{i+1}, \cdots, x_n)$$

where $1 \leq i \leq n$. Notice we have $12 - n$ enrichment maps.

Given $(x_1, x_2, \cdots, x_i, x_{i+1}, \cdots, x_n) \in \mathcal{T}^n$, the intervals $x_i$ and $x_{i+1}$ can be joined together through the removal of one note in the chord. A **reduction** $\pi_i$ is a map from $\mathcal{T}^n$ to $\mathcal{T}^{n-1}$ defined by:

$$\pi_i(x_1, x_2, \cdots, x_i, x_{i+1}, \cdots, x_n) = (x_1, x_2, \cdots, x_i + x_{i+1}, \cdots, x_n)$$

for $1 \leq i < n$,

and

$$\pi_n(x_1, x_2, \cdots, x_i, x_{i+1}, \cdots, x_n) = (x_n + x_1, x_2, \cdots, x_i + x_{i+1}, \cdots, x_{n-1})$$

## Examples

$$f_3^1(2, 2, 4, 4) = (2, 2, 1, 3, 4)$$

$$\pi_4(1, 2, 1, 1, 2, 5) = (1, 2, 1, 3, 5)$$

Enrichments and reductions are inverses of each other in the following manner: Notice that $\pi_i \circ f_i^j = Id^n$, for every $1 \leq j \leq x_i - 1$, where $Id^n$ denotes the identity map in $\mathcal{T}^n$.

Conversely, given $(x_1, x_2, \cdots, x_i, \cdots, x_n) \in \mathcal{T}^n$, by setting the value of $j = x_i$ we have:

$$f_i^j \circ \pi_i(x_1, \cdots, x_n) = f_i^{x_i}(x_1, \cdots, x_i + x_{i+1}, \cdots, x_n) = (x_1, \cdots, x_n)$$

In order to avoid negative sub-indexes, let's adopt from now onwards the following natural identities based on modulo $n$:

$$f_{n+i}^j = f_i^j \qquad \text{and} \qquad \pi_{n+i} = \pi_i$$

We are going to derive properties regarding the effect of the permutations $\sigma$ on enrichments and reductions:

$$\sigma \circ f_i^j(x_1,\cdots,x_i,\cdots,x_n) = \sigma(x_1,\cdots,x_{i-1},j,x_i-j,x_{i+1},\cdots,x_n) = (x_2,\cdots,x_{i-1},j,x_i-j,\cdots,x_n,x_1)$$

Also,

$$f_i^j \circ \sigma(x_1,\cdots,x_i,\cdots,x_n) = f_{i-1}^j(x_2,\cdots,xi,\cdots,x_n) = (x_2,\cdots,x_{i-1},j,x_i-j,\cdots,x_n,x_1)$$

Thus,

$$\sigma \circ f_i^j = f_{i-1}^j \circ \sigma \quad , \text{and} \quad \sigma^2 \circ f_i^j = \sigma \circ \sigma \circ f_i^j = \sigma \circ f_{i-1}^j = f_{i-2}^j \circ \sigma^2$$

And inductively:

$$\boxed{\sigma^k \circ f_i^j = f_{i-k}^j \circ \sigma^k \quad , \text{for } 1 \le k < i.} \quad (3.1)$$

By (3.1), the following diagram is commutative:

$$
\begin{array}{ccccc}
\mathcal{T}^n & \xrightarrow{\ \sigma\ } & \mathcal{T}^n & \xrightarrow{\ \sigma\ } & \mathcal{T}^n \\
f_i^j \downarrow & & f_{i-1}^j \downarrow & & f_{i-2}^j \downarrow \qquad i \ge 3 \\
\mathcal{T}^{n+1} & \xrightarrow{\ \sigma\ } & \mathcal{T}^{n+1} & \xrightarrow{\ \sigma\ } & \mathcal{T}^{n+1}
\end{array}
$$

If $k > i$, then

$$\sigma^k \circ f_i^j(x_1,\cdots,x_i,\cdots,x_k,\cdots,x_n) = \sigma^k(x_1,\cdots,j,x_i-j,\cdots,x_k,\cdots,x_n)$$

$$= (x_k,\cdots,x_n,x_1,\cdots,x_{i-1},j,x_i-j,x_{i+1},\cdots,x_{k-1})$$

and

$$f_{n-k+i+1}^j \circ \sigma^k(x_1,\cdots,x_i,\cdots,x_k,\cdots,x_n) = f_{n-k+i}^j(x_k,\cdots,x_n,x_1,\cdots,x_i,\cdots,x_{k-1})$$

$$= (x_k,\cdots,x_n,x_1,\cdots,x_{i-1},j,x_i-j,x_{i+1},\cdots,x_{k-1})$$

Therefore

$$\boxed{\sigma^k \circ f_i^j = f_{n-k+i+1}^j \circ \sigma^{k-1} \quad \text{for } k > i.} \quad (3.1A)$$

It is immediate to check that $\sigma^k \circ f_k^j$ does not commute. This does not present the slightest inconvenience since we can always find a permutation of power $\sigma^k$ such that $\sigma^k \circ f_i^j$ commutes if $k \ne i$.

**Example** Case where $k < i$.

$$\sigma^2 \circ f_4^3(1, 2, 2, 4, 3) = \sigma^2(1, 2, 2, 3, 1, 3) = (2, 3, 1, 3, 1, 2)$$

and

$$f_2^3 \circ \sigma^2(1, 2, 2, 4, 3) = f_2^3(2, 4, 3, 1, 2) = (2, 3, 1, 3, 1, 2)$$


**Example** Case where $k > i$.

$$\sigma^4 \circ f_3^1(1, 2, 4, 5) = \sigma^4(1, 2, 1, 3, 5) = (5, 1, 2, 1, 3)$$

and

$$f_4^1 \circ \sigma^3(1, 2, 4, 5) = f_4^1(5, 1, 2, 4) = (5, 1, 2, 1, 3)$$

The reductions share similar properties.

$$\sigma \circ \pi_i(x_1, \cdots, x_i, x_{i+1}, \cdots, x_n) = \sigma(x_1, \cdots, x_i + x_{i+1}, \cdots, x_n) = (x_2, \cdots, x_i + x_{i+1}, \cdots, x_n, x_1)$$

and

$$\pi_{i-1} \circ \sigma(x_1, \cdots, x_i, x_{i+1}, \cdots, x_n) = \pi_{i-1}(x_2, \cdots, x_i + x_{i+1}, \cdots, x_n, x_1) = (x_2, \cdots, x_i + x_{i+1}, \cdots, x_n, x_1)$$

and hence $\sigma \circ \pi_i = \pi_{i-1} \circ \sigma$.

Now, $\sigma^2 \circ \pi_i = \sigma \circ \sigma \circ \pi_i = \sigma \circ \pi_{i-1} \circ \sigma = \pi_{i-2} \circ \sigma^2$.

And inductively,

$$\boxed{\sigma^k \circ \pi_i = \pi_{i-k} \circ \sigma^k \qquad \text{for } 1 \leq k < i.} \qquad (3.2)$$

The above identity makes the following diagram commute:

$$
\begin{array}{ccccc}
\mathcal{T}^n & \xrightarrow{\sigma} & \mathcal{T}^n & \xrightarrow{\sigma} & \mathcal{T}^n \\
\pi_i \downarrow & & \pi_{i-1} \downarrow & & \pi_{i-2} \downarrow \qquad i \geq 3 \\
\mathcal{T}^{n-1} & \xrightarrow{\sigma} & \mathcal{T}^{n-1} & \xrightarrow{\sigma} & \mathcal{T}^{n-1}
\end{array}
$$

Now, if $k \geq i$

$$\sigma^k \circ \pi_i(x_1, \cdots, x_i, x_{i+1}, \cdots, x_k, \cdots, x_n) = \sigma^k(x_1, \cdots, x_i + x_{i+1}, \cdots, x_k, \cdots, x_n)$$

$$= (x_{k+2}, \cdots, x_n, x_1, \cdots, x_i + x_{i+1}, \cdots, x_k, x_{k+1})$$

and

$$\pi_{n-k+i-1} \circ \sigma^{k+1}(x_1, \cdots, x_i, x_{i+1}, \cdots, x_k, \cdots, x_n) = \pi_{n-k+i-1}(x_{k+2}, \cdots, x_n, x_1, \cdots, x_i, x_{i+1}, \cdots, x_k, x_{k+1})$$

$$= (x_{k+2}, \cdots, x_n, x_1, \cdots, x_i + x_{i+1}, \cdots, x_k, x_{k+1})$$

Thus,

$$\boxed{\sigma^k \circ \pi_i = \pi_{n-k+i-1} \circ \sigma^{k+1} \quad \text{for } k \geq i.} \qquad (3.2\text{A})$$

Note: Actually the case $k = i - 1$ can be solved by both (3.2) and (3.2A) since $\pi_n \circ \sigma^k = \pi_1 \circ \sigma^{k-1}$.

## Inversion properties of enrichments and reductions

Consider inversion $I$ and enrichment $f_i^j$ acting on a chord $P = (x_1, x_2, \cdots, x_i, \cdots, x_n)$.

$$I \circ f_i^j(x_1, x_2, \cdots, x_i, \cdots, x_n) = I(x_1, \cdots, x_{i-1}, j, x_i - j, x_{i+1}, \cdots, x_n) = (x_n, \cdots, x_{i+1}, x_i - j, j, x_{i-1}, \cdots, x_1)$$

$$f_{n-i+1}^{x_i - j} \circ I(x_1, x_2, \cdots, x_i, \cdots, x_n) = f_{n-i+1}^{x_i - j}(x_n, \cdots, x_i, \cdots, x_1) = (x_n, \cdots, x_{i+1}, x_i - j, j, x_{i-1}, \cdots, x_1)$$

and thus,

$$\boxed{I \circ f_i^j = f_{n-i+1}^{x_i - j} \circ I} \qquad (3.3)$$

## Example

$$I \circ f_3^2(3, 1, 3, 5) = I(3, 1, 2, 1, 5) = (5, 1, 2, 1, 3)$$

and

$$f_2^1 \circ I(3, 1, 3, 5) = f_2^1(5, 3, 1, 3) = (5, 1, 2, 1, 3)$$

If chord $P$ and $Q$ in $\mathcal{T}^n$ are co-$I$-symmetric, i.e. $(\overline{P}) = I(\overline{Q})$, then

$$\sigma^k(P) = I(Q)$$

where $k$ is the inversion exponent of $P$ with respect to $Q$.

Now, from (3,3), we obtain:

$$I \circ f_i^j(Q) \;=\; f_{n-i+1}^{x_i-j} \circ I(Q) \;=\; f_{n-i+1}^{x_i-j} \circ \sigma^k(P) \;=\; \begin{cases} \sigma^k \circ f_{n-i+1+k}^{x_i-j}(P), & k+1 \le i \le n \quad\quad (3.1) \\[2mm] \sigma^{k+1} \circ f_{n-i+1+k}^{x_i-j}(P), & 1 \le i \le k \quad\quad\; (3.1A) \end{cases}$$

whence $\boxed{f_i^j(Q) \text{ and } f_{n-i+1+k}^{x_i-j}(P) \text{ are co-}I\text{-symmetric, for all } 1 \le i \le n}$. (3.4)

**Example**: $P = (1, 3, 2, 6)$ and $Q = (3, 1, 6, 2)$

The inversion exponent of $P$ with respect to $Q$ is 2. So by (3.4), we have $f_2^1(P)$ and $f_1^2(Q)$ are co-$I$-symmetric. In fact,

$$f_2^1(1, 3, 2, 6) = (1, 1, 2, 2, 6) \quad \text{and} \quad f_1^2(3, 1, 6, 2) = (2, 1, 1, 6, 2)$$

and thus $\overline{I(1, 1, 2, 2, 6)} = \overline{(2, 1, 1, 6, 2)}$

Also,

$$f_4^1(1, 3, 2, 6) = (1, 3, 2, 1, 5)$$

$$f_3^5(3, 1, 6, 2) = (3, 1, 5, 1, 1)$$

and thus $\overline{I(1, 3, 2, 1, 5)} = \overline{(3, 1, 5, 1, 2)}$

The following pairs are also co-$I$-symmetric:

$f_2^2(P)$ and $f_1^1(Q)$,
$f_3^1(P)$ and $f_4^1(Q)$,
$f_4^2(P)$ and $f_3^4(Q)$,
$f_4^3(P)$ and $f_3^3(Q)$,
$f_4^4(P)$ and $f_3^2(Q)$,
$f_4^5(P)$ and $f_3^1(Q)$.

Now, if $P$ is $I$-symmetric, $P = Q$ above, and $f_i^j(P)$ and $f_{n-i+1+k}^{x_i-j}(P)$ are co-$I$-symmetric.

Notice, furthermore if $j = x_i - j$ and $i = n - i + 1 + k$, that is,

$$\boxed{\text{If } j = \frac{x_i}{2} \text{ and } i = \frac{n+k+1}{2} \text{ then } f_i^j(P) \text{ is } I\text{-symmetric.}} \quad (3.5)$$

**Example**: Consider the $I$-symmetric 4-chord $P = (1, 4, 1, 6)$, where $I(P) = (6, 1, 4, 1)$ and $k = 3$. By (3.5) $f_2^2(P)$ and $f_4^3(P)$ are $I$-symmetric. In fact,

$$f_2^2(1, 4, 1, 6) = (1, 2, 2, 1, 6)$$

and

$$f_4^3(1, 4, 1, 6) = (1, 4, 1, 3, 3).$$

The composition of inversions and reductions gives us:

$$\pi_i \circ I(x_1, \cdots, x_n) = \pi_i(x_n, \cdots, x_1) = (x_n, x_{n-1}, \cdots, x_{n-i+1} + x_{n-i}, \cdots, x_1)$$

$$I \circ \pi_{n-i}(x_1, \cdots, x_n) = I(x_1, x_2, \cdots, x_{n-i} + x_{n-i+1}, \cdots, x_n) = (x_n, x_{n-1}, \cdots, x_{n-i} + x_{n-i+1}, \cdots, x_1)$$

and

$$\boxed{\pi_i \circ I = I \circ \pi_{n-i} \quad , \text{ for } i < n} \quad (3.6)$$

Also

$$\sigma \circ \pi_n \circ I(x_1, \cdots, x_n) = \sigma \circ \pi_n(x_n, \cdots, x_1) = \sigma(x_1 + x_n, x_{n-1}, \cdots, x_2) = (x_{n-1}, \cdots, x_2, x_1 + x_n)$$

$$I \circ \pi_n(x_1, \cdots, x_n) = I(x_n + x_1, x_2 \cdots, x_{n-1}) = (x_{n-1}, \cdots, x_2, x_n + x_1)$$

and

$$\boxed{\sigma \circ \pi_n \circ I = I \circ \pi_n \quad \text{or} \quad \pi_n \circ I = \sigma_{n-1} \circ I \circ \pi_n} \quad (3.7)$$

**Examples**:

$$\pi_3 \circ I(1, 3, 2, 2, 4) = \pi_3(4, 2, 2, 3, 1) = (4, 2, 5, 1)$$

and

$$I \circ \pi_2(1, 3, 2, 2, 4) = I(1, 5, 2, 4) = (4, 2, 5, 1)$$

Now,

$$\pi_5 \circ I(1, 3, 2, 2, 4) = \pi_5(4, 2, 2, 3, 1) = (5, 2, 2, 3)$$

and

$$I \circ \pi_5(1, 3, 2, 2, 4) = I(5, 3, 2, 2) = (2, 2, 3, 5)$$

hence

$$\sigma \circ \pi_5 \circ I = I \circ \pi_5$$

Given two co-*I*-symmetric chords $P$ and $Q$ with inversion exponent $k$,

$$\sigma^k(P) = I(Q)$$

$$I \circ \pi_i(Q) \overset{(3.6)}{=} \pi_{n-i} \circ I(Q) = \pi_{n-i} \circ \sigma^k(P) = \begin{cases} \sigma^k \circ \pi_{n-i+k}, & k+1 \le i < n \qquad (3.2) \\[2ex] \sigma^{k-1} \circ \pi_{n-i+k}(P), & 1 \le i \le k \qquad (3.2A) \end{cases}$$

Also

$$I \circ \pi_n(Q) \overset{(3.7)}{=} \sigma \circ \pi_n \circ I(Q) = \sigma \circ \pi_n \circ \sigma_k(P) \overset{(3.2A)}{=} \sigma \circ \sigma^{k-1} \circ \pi_{n+k}(P) = \sigma^k \circ \pi_{n+k}(P)$$

and

$$\pi_n(Q) \text{ and } \pi_{n+k}(P) = \pi_k(P) \text{ are co-}I\text{-symmetric.}$$

Therefore if $1 \le i \le n$

$$\boxed{\pi_i(Q) \text{ and } \pi_{n-i+k}(P) \text{ are co-}I\text{-symmetric.}} \qquad (3.8)$$

**Example**

Let $P = (2, 1, 4, 5)$ and $Q = (1, 2, 5, 4)$. Since $\sigma^2 (2, 3, 4, 5) = I(Q) = (4, 5, 2, 1)$, we have $k = 2$.

From (3.8),

$\pi_1 (1, 2, 5, 4) = (3, 5, 4)$ is co-*I*-symmetric with $\pi_1 (2, 1, 4, 5) = (3, 4, 5)$.

$\pi_2 (1, 2, 5, 4) = (1, 7, 4)$ is co-*I*-symmetric with $\pi_4 (2, 1, 4, 5) = (7, 1, 4)$.

$\pi_3 (1, 2, 5, 4) = (1, 2, 9)$ is co-*I*-symmetric with $\pi_3 (2, 1, 4, 5) = (2, 1, 9)$.

$\pi_4 (1, 2, 5, 4) = (5, 2, 5)$ is co-*I*-symmetric with $\pi_2 (2, 1, 4, 5) = (2, 5, 5)$.

Now if $P$ is *I*-symmetric, $P = Q$ above and $\pi_i(P)$ and $\pi_{n-i+k}(P)$ are co-*I*-symmetric. Notice furthermore that if $i = n - i + k$, or

$$\boxed{\text{if } i = \frac{n+k}{2} \text{ or } i = \frac{k}{2} \text{, then } \pi_i(P) \text{ is } I\text{-symmetric.}} \quad (3.9)$$

## Example

Let $P = (1, 1, 4, 2, 4)$. Since $I(P) = (4, 2, 4, 1, 1)$, then $k = 2$. Since $i = \dfrac{5+2}{2}$ is not a natural

number, we take $i = \dfrac{2}{2} = 1$.

By (3.9), we have that $\pi_1 (1, 1, 4, 2, 4) = (2, 4, 2, 4)$ is $I$-symmetric.

## Enrichments and Reductions: Metric Properties

Given two chords $P = (x_1, \cdots, x_n)$ and $Q = (y_1, \cdots, y_n)$ and their enrichments $f_i^j (P)$ and $f_i^j (Q)$,

$$f_i^j (P) = (x_1, \cdots, x_{i-1}, j, x_i - j, x_{i+1}, \cdots, x_n) \text{ and } f_i^j (Q) = (y_1, \cdots, y_{i-1}, j, y_i - j, y_{i+1}, \cdots, y_n).$$

Since $\left((x_i - j) - (y_i - j)\right)^2 = (x_i - y_i)^2$ we have:

$$\delta\!\left(f_i^j (P), f_i^j (Q)\right) = \delta(P, Q) \qquad (3.10)$$

If $n$ divides 12 and we take $Q$ to be the barycentric chord $B$ in $\mathcal{T}^n$, then from (3.10) we immediately obtain:

$$\text{Index } (P) = \delta\!\left(f_i^j (P), f_i^j (B)\right) \qquad (3.11)$$

for all values of $i$ and $j$ where $f_i^j$ are well defined.

## Examples

Given a chord $P = (1, 2, 2, 7)$ in $\mathcal{T}^4$ and the barycenter $B = (3, 3, 3, 3)$,

$$\text{Index } P = \sqrt{11}$$

$$\delta\!\left(f_4^2 (P), f_4^2 (B)\right) = \delta\!\left((1, 2, 2, 2, 5), (3, 3, 3, 2, 1)\right) = \sqrt{11}$$

$$\delta\!\left(f_2^1 (P), f_2^1 (B)\right) = \delta\!\left((1, 1, 1, 2, 7), (3, 1, 2, 3, 3)\right) = \sqrt{11}$$

Illustrating formula (3.11).

Given a chord $P = (x_1, \cdots, x_n)$ and two of its enrichments

$$f_i^j (x_1, \cdots, x_n) = (x_1, \cdots, x_{i-1}, j, x_i - j, x_{i+1}, \cdots, x_n)$$

and

$$f_i^k (x_1, \cdots, x_n) = (x_1, \cdots, x_{i-1}, k, x_i - k, x_{i+1}, \cdots, x_n)$$

then

$$\text{Index } f_i^k(P) \geq \text{Index } f_i^j(P) \quad \Leftrightarrow \quad (k-b)^2 + (x_i - k - b)^2 \geq (j-b)^2 + (x_i - j - b)^2$$

where $b = \dfrac{12}{n+1}$

This inequality reduces to $2k^2 - 2x_i k \geq 2j^2 - 2x_i j$, or $k(k - x_i) \geq j(j - x_i)$, so we have:

$$\boxed{\text{Index } f_i^k(P) \geq \text{Index } f_i^j(P) \quad \Leftrightarrow \quad k(x_i - k) \leq j(x_i - j)} \qquad (3.12)$$

Notice that the maximum evenness (consonance) of $f_i^j(P)$ is obtained when $j(x_i - j)$ is

maximum. This is achieved when $j = x_i - j$, or $j = \dfrac{x_i}{2}$.

Consider:

$$F(j) = j(x_i - j) = -j^2 + j \cdot x_i \ ,$$

with derivatives:

$$F'(j) = -2j + x_i$$

$$F''(j) = -2$$

The maximum is achieved when $F'(j) = 0$.

Therefore the equal or almost equal partition of the interval $x_i$ produces more consonance.

From (3.12) we obtain the equality

$$\text{Index } f_i^k(P) = \text{Index } f_i^j(P) \quad \Leftrightarrow \quad k(x_i - k) = j(x_i - j)$$

which derives the following result

$$\boxed{\text{Index } f_i^j(P) = \text{Index } f_i^{x_i - j}(P)}$$

**Example**

Consider $P = (1, 1, 2, 2, 6)$ in $\mathcal{T}^5$.

$f_5^1(P) = (1, 1, 2, 2, 1, 5)$  $\qquad\qquad$ $f_5^4(P) = (1, 1, 2, 2, 4, 2)$

$f_5^2(P) = (1, 1, 2, 2, 2, 4)$  $\qquad\qquad$ $f_5^5(P) = (1, 1, 2, 2, 5, 1)$

$f_5^3(P) = (1, 1, 2, 2, 3, 3)$

Index $f_5^1(P)$ = Index $f_5^5(P)$ = $\sqrt{6}$

Index $f_5^2(P)$ = Index $f_5^4(P)$ = $\sqrt{3}$

Index $f_5^3(P)$ = $\sqrt{2}$ , which gives us the

maximum evenness (consonance).

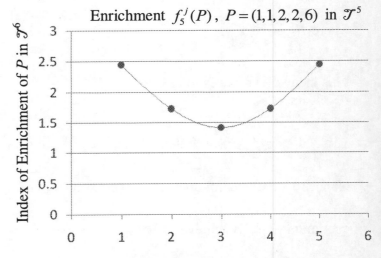

Enrichment $f_5^j(P)$, $P = (1,1,2,2,6)$ in $\mathcal{T}^5$

Index of Enrichment of $P$ in $\mathcal{T}^6$

Value of $j$ partitioning interval $x_5=6$

Given two chords $P = (x_1, \cdots, x_n)$ and $Q = (y_1, \cdots, y_n)$ in $\mathcal{T}^n$, and their reductions $\pi_i(P)$ and $\pi_i(Q)$, we are going to compare the distances $\delta(P,Q)$ and $\delta(\pi_i(P), \pi_i(Q))$.

$$2(\delta(P,Q))^2 = \sum_{i=1}^{n}(x_i - y_i)^2$$

and

$$2(\delta(\pi_i(P), \pi_i(Q)))^2 = \sum_{k=1}^{i-1}(x_k - y_k)^2 + \left((x_i + x_{i+1}) - (y_i + y_{i+1})\right)^2 + \sum_{k=i+2}^{n}(x_k - y_k)^2$$

Since

$$\left((x_i + x_{i+1}) - (y_i + y_{i+1})\right)^2 = \left((x_i - y_i) + (x_{i+1} - y_{i+1})\right)^2$$

$$= (x_i - y_i)^2 + (x_{i+1} - y_{i+1})^2 + 2(x_i - y_i)(x_{i+1} - y_{i+1})$$

After the appropriate cancellation of terms in the inequality, we arrive at:

$$\delta(P,Q) > \delta(\pi_i(P), \pi_i(Q)) \quad \Leftrightarrow \quad (x_i - y_i)(x_{i+1} - y_{i+1}) < 0$$

$$\Leftrightarrow \quad \text{sign}(x_i - y_i) \neq \text{sign}(x_{i+1} - y_{i+1})$$

Similarly,

$$\delta(P,Q) < \delta(\pi_i(P), \pi_i(Q)) \quad \Leftrightarrow \quad \text{sign}(x_i - y_i) = \text{sign}(x_{i+1} - y_{i+1})$$

and

$$\boxed{\delta(P,Q) = \delta(\pi_i(P), \pi_i(Q)) \quad \Leftrightarrow \quad x_i = y_i, \text{ or } x_{i+1} = y_{i+1}.} \qquad (3.13)$$

### Example

Given chords $P = (2, 3, 2, 1, 4)$ and $Q = (2, 2, 2, 1, 5)$ with distance $\delta(P,Q) = 1$.
Since (3.13) is satisfied by every reduction $\pi_i(P)$ and $\pi_i(Q)$, $1 \leq i \leq 6$, all these reductions are at a distance one of each other.

We are going to obtain a remarkable result about the evenness index of different reductions.
Given a chord $P = (x_1, \cdots, x_n)$ in $\mathscr{T}^n$,

$$\text{Index } \pi_i(P) \geq \text{Index } \pi_j(P)$$

$$\Leftrightarrow (x_1 - b)^2 + \cdots + (x_i + x_{i+1} - b)^2 + \cdots + (x_n - b)^2 \geq (x_1 - b)^2 + \cdots + (x_j + x_{j+1} - b)^2 + \cdots + (x_n - b)^2$$

where $b = \dfrac{12}{(n-1)}$, and $(b, b, \cdots, b)$ is the barycenter in $\mathscr{T}^{n-1}$.

After pertinent cancellations, the inequality reduces to

$$(x_i + x_{i+1} - b)^2 + (x_j - b)^2 + (x_{j+1} - b)^2 \geq (x_j + x_{j+1} - b)^2 + (x_i - b)^2 + (x_{i+1} - b)^2$$

which reduces itself to:

$$x_i \cdot x_{i+1} \geq x_j \cdot x_{j+1}$$

Since

$$\boxed{\text{Index } \pi_i(P) \geq \text{Index } \pi_j(P) \quad \Leftrightarrow \quad x_i \cdot x_{i+1} \geq x_j \cdot x_{j+1}} \qquad (3.14)$$

The maximum evenness (consonance) is achieved in the reduction corresponding to the minimum product of adjacent intervals.

**Example**

Consider $P = (1, 2, 2, 3, 4)$ the 9th with a 7th major chord.

The products of adjacent intervals:

$$x_1 \cdot x_2 = 2$$
$$x_2 \cdot x_3 = 4$$
$$x_3 \cdot x_4 = 6$$
$$x_4 \cdot x_5 = 12$$
$$x_5 \cdot x_1 = 4$$

If we want to maintain the chord character of the original chord in the reduction, then we consider the one with the highest evenness index, namely $\pi_4(P)$ which is associated to the highest product $x_4 \cdot x_5 = 12$. Notice that $\pi_4(P) = (1, 2, 2, 7)$ eliminates the 5th of the chord, keeping the 9th and the 7th major. On the opposite side of the spectrum, if we consider the most consonant reduction $\pi_1(P) = (3, 2, 3, 4)$, i.e. a minor 7th or major 6th, the character of the original chord P is already lost.

Notice that the most consonant reductions of $\pi_1(P)$ gives us:

$$\pi_1(\pi_1(P)) = (5, 3, 4) \quad \text{minor triad}$$
$$\pi_2(\pi_1(P)) = (3, 5, 4) \quad \text{major triad}$$

# 4. Degenerated Chords

The case $\mathcal{T}^4$ presents chords with special properties that exemplify the structure of chords in interval space. Not only does $\mathcal{T}^4$ provide a clear 3-dimensional outlook of different chords, but also reveals the intrinsic symmetries associated with the cyclic sequence of intervals, allowing for insight into chords with exceptional properties. While most chords in $\mathcal{T}^4$ are represented by tetrahedral orbitals, some chords collapse into two diametrically opposed points (polar chords), and some chords collapse onto a planar square with vertices in a maximum circle (equatorial chords). These types of chords are part of a broader category that we will define later as degenerated chords. (See Figures 4.1a and 4.1b.)

**Examples**

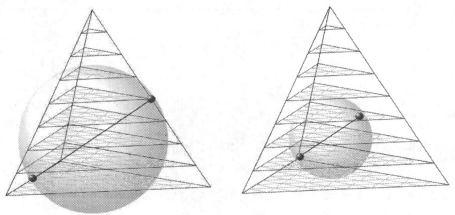

**Figure 4.1a** Polar chords (1, 5, 1, 5) and (2, 4, 2, 4)

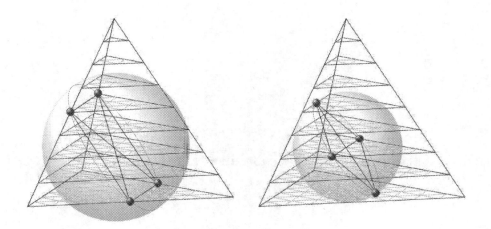

**Figure 4.1b** Equatorial chords (1, 1, 5, 5) and (3, 1, 3, 5)

A chord in $\mathscr{T}^n$ is a **polar chord** if and only if $\widehat{\sigma_2}(P) = 0$. Equatorial chords have a more complicated characterization, since they depend on the cardinality of the chord and the dimension of the hyper-sphere where these chords are immersed. In order to classify these degenerated properties in higher dimensions we introduce two different types of cyclic permutation matrices associated with the components of a chord. Given a chord $P$ in $\mathscr{T}^n$, we define two maps from $\mathscr{T}^n$ into $\mathscr{M}^n(\mathbb{Q})$, the linear space of square $n \times n$ matrices with rational entries.

The map $M_H$ sends chord $P$ to the matrix in $\mathscr{M}^n(\mathbb{Q})$ with rows:

$$P, \sigma(P), \sigma^2(P), \cdots, \sigma^{n-1}(P)$$

that is,

$$M_H(P) = \begin{pmatrix} x_1 & x_2 & x_3 & \cdots & x_n \\ x_2 & x_3 & x_4 & \cdots & x_1 \\ x_3 & x_4 & x_5 & \cdots & x_2 \\ \vdots & \vdots & \vdots & \ddots & \vdots \\ x_n & x_1 & x_2 & \cdots & x_{n-1} \end{pmatrix}$$

This type of cyclic permutation matrices are called **Hankel matrices**.

The map $M_C$ sends chord $P$ to the matrix in $\mathscr{M}^n(\mathbb{Q})$ with rows:

$$P, \sigma^{-1}(P), \sigma^{-2}(P), \cdots, \sigma^{-n+1}(P)$$

that is,

$$M_C(P) = \begin{pmatrix} x_1 & x_2 & x_3 & \cdots & x_n \\ x_n & x_1 & x_2 & \cdots & x_{n-1} \\ x_{n-1} & x_n & x_1 & \cdots & x_{n-2} \\ \vdots & \vdots & \vdots & \ddots & \vdots \\ x_2 & x_3 & x_4 & \cdots & x_1 \end{pmatrix}$$

This type of cyclic permutation matrices are called **circulant matrices** and they are very fruitful in different areas of mathematics.

**Examples**

Given $P = (1, 2, 4, 5)$ in $\mathcal{T}^4$,

$$M_H(P) = \begin{pmatrix} 1 & 2 & 4 & 5 \\ 2 & 4 & 5 & 1 \\ 4 & 5 & 1 & 2 \\ 5 & 1 & 2 & 4 \end{pmatrix} \quad , \quad M_C(P) = \begin{pmatrix} 1 & 2 & 4 & 5 \\ 5 & 1 & 2 & 4 \\ 4 & 5 & 1 & 2 \\ 2 & 4 & 5 & 1 \end{pmatrix}$$

These matrices are extremely closely related. For example,

$$\left| \det\left(M_H(P)\right) \right| = \left| \det\left(M_C(P)\right) \right|$$

Moreover, let $\lambda_0^H, \lambda_1^H, \cdots, \lambda_{n-1}^H$ represent the eigenvalues of $M_H(P)$, and let $\lambda_0^C, \lambda_1^C, \cdots, \lambda_{n-1}^C$ represent the eigenvalues of $M_C(P)$, then we have the following properties:

i) $\quad \lambda_0^H = \lambda_0^C = 12$

ii) $\quad \lambda_{n/2}^H = \lambda_{n/2}^C = x_1 - x_2 + x_3 - x_4 + \cdots + x_{n-1} - x_n \qquad$ if $n$ is even

iii) $\quad \left| \lambda_i^H \right| = \left| \lambda_i^C \right| \quad$ , where $i = 1, \cdots, n-1$

Since the values of the Hankel matrix and its eigenvalues for a given chord do not change (up to the sign) with the values of the Hankel matrix and the eigenvalues of the inversion of the chord and its different permutations, it is natural to extend these matricial properties to $\mathcal{T}^n/\Sigma^n$ and $\mathcal{T}^n/\Phi^n$.

Although the complex eigenvalues of the circulant matrix of a chord change significantly with permutations and inversions, the unique properties of these circulant matrices make them ideal to manipulate the corresponding eigenvalues, eigenvectors, and their relationship to the orbital meshes $\widehat{\sigma_k}$.

The circulant matrices have the advantageous property that their eigenvalues are linear combinations of multiples of $n$th roots of unity and their eigenvectors are the same for a fixed $n$. (Gray [2]). Namely, if $\varepsilon = e^{\frac{2\pi i}{n}}$ is the primitive $n^{\text{th}}$ root of unity. The eigenvalues of $M_C(P)$ are then expressed as:

$$\lambda_\ell = \sum_{k=1}^{n} x_k \varepsilon^{(k-1)\ell} \qquad , \ell = 0, 1, 2, \cdots, n-1$$

where $P = (x_1, x_2, \cdots, x_n)$. Note that $\det\left(M_C(P)\right) = \prod_{\ell=0}^{n-1} \lambda_\ell$. The corresponding normalized eigenvectors are:

$$\vec{v}_\ell = \frac{1}{\sqrt{n}}\left(1, \varepsilon^\ell, \varepsilon^{2\ell}, \cdots, \varepsilon^{(n-1)\ell}\right) \qquad , \ell = 0, 1, 2, \cdots, n-1$$

where these eigenvectors do not depend on $P$.

The $n \times n$ matrix with $\vec{v}_\ell$ as column vectors:

$$F = \left(\vec{v}_0, \vec{v}_1, \vec{v}_2, \cdots, \vec{v}_{n-1}\right) = \frac{1}{\sqrt{n}}\begin{pmatrix} 1 & 1 & 1 & 1 & \cdots & 1 \\ 1 & \varepsilon & \varepsilon^2 & \varepsilon^3 & \cdots & \varepsilon^{n-1} \\ 1 & \varepsilon^2 & \varepsilon^4 & \varepsilon^6 & \cdots & \varepsilon^{2(n-1)} \\ 1 & \varepsilon^3 & \varepsilon^6 & \varepsilon^9 & \cdots & \varepsilon^{3(n-1)} \\ \vdots & \vdots & \vdots & \vdots & \ddots & \vdots \\ 1 & \varepsilon^{n-1} & \varepsilon^{2(n-1)} & \varepsilon^{3(n-1)} & \cdots & \varepsilon^{(n-1)(n-1)} \end{pmatrix}$$

is the Fourier matrix that takes $A = M_C(P)$ into a diagonal matrix $D = Diag\left(\lambda_0, \lambda_1, \cdots, \lambda_{n-1}\right)$, such that

$$F^{-1}A\,F = D$$

We are going to use this machinery of circulant matrices to express the orbital meshes and the index of dissonance in terms of the eigenvalues.

Due to the orthogonality of the complex exponentials, the Fourier matrix $F$ is unitary and $F^{-1} = F^*$ (the conjugate-transpose of $F$). Now $F^* A = D F^*$, and since unitary matrices preserve the inner-product and therefore the Euclidian metric $\| \cdot \|_e$ :

$$2\left(\widehat{\sigma}_k(P)\right)^2 = \left\| A(\vec{e}_1 - \vec{e}_{k+1}) \right\|_e^2 = \left\| F^* A(\vec{e}_1 - \vec{e}_{k+1}) \right\|_e^2$$

Observe that the 1st column in $F^*$ is

$$F^*(\vec{e}_1) = \frac{1}{\sqrt{n}} \left(1, 1, \cdots, 1\right)^T$$

and the $k + 1$ column is

$$F^*(\vec{e}_{k+1}) = \frac{1}{\sqrt{n}} \left(1, \bar{\varepsilon}^k, \bar{\varepsilon}^{2k}, \cdots, \bar{\varepsilon}^{(n-1)k}\right)^T \qquad \text{where } \bar{\varepsilon} = e^{-\frac{2\pi i}{n}}$$

Therefore,

$$\left\| D F^*(\vec{e}_1 - \vec{e}_{k+1}) \right\|_e^2 = \frac{1}{n} \sum_{j=1}^{n-1} \left|\lambda_j\right|^2 \left|1 - \bar{\varepsilon}^{jk}\right|^2$$

and hence,

$$\boxed{\left(\widehat{\sigma}_k(P)\right)^2 = \frac{1}{2n} \sum_{j=1}^{n-1} \left|\lambda_j\right|^2 \left|1 - \varepsilon^{jk}\right|^2} \qquad (4.1) \qquad \text{since } \left|1 - \varepsilon^{jk}\right|^2 = \left|1 - \bar{\varepsilon}^{jk}\right|^2$$

It is easy to check that

$$\lambda_j = \overline{\lambda_{n-j}} \qquad \text{and so} \qquad \left|\lambda_j\right|^2 = \left|\lambda_{n-j}\right|^2 \quad, \ j = 1, \cdots, n-1$$

Also,

$$\left|1 - \varepsilon^{kj}\right|^2 = \left|1 - \varepsilon^{k(n-j)}\right|^2 \qquad\qquad j = 1, \cdots, n-1$$

Now take $k = \left[\!\left[ \dfrac{12}{n} \right]\!\right]$ and the column matrices

$$\left(\widehat{\sigma^2}\right) = \left(\widehat{\sigma_1}^2, \widehat{\sigma_2}^2, \cdots, \widehat{\sigma_k}^2\right)^T$$

$$\left(\lambda^2\right) = \left(|\lambda_1|^2, |\lambda_2|^2, \cdots, |\lambda_k|^2\right)^T$$

We construct the passage matrix $P_n = \left(p_{ij}\right)$ as

$$p_{ij} = \left|1 - \varepsilon^{ij}\right|^2 \quad \text{for } i = 1, \cdots, k \text{ and } j = 1, \cdots, k-1$$

and

$$p_{ik} = \begin{cases} \left|1 - \varepsilon^{ik}\right|^2 & \text{if } n \text{ is odd} \\[2mm] \dfrac{\left|1 - \varepsilon^{ik}\right|^2}{2} & \text{if } n \text{ is even} \end{cases}$$

From the Equality (4.1) we immediately obtain

$$\boxed{\left(\widehat{\sigma^2}\right) = \frac{1}{n} P_n \left(\lambda^2\right)}$$

The matrices $P_n$ are non-singular, and then

$$\boxed{\left(\lambda^2\right) = n P_n^{-1} \left(\lambda^2\right)}$$

where $P_n^{-1} = \left(q_{ij}\right)$ are defined by

$$q_{ij} = \frac{p_{ij} - 2}{n} = \frac{-\left(\varepsilon^{ij} + \overline{\varepsilon}^{ij}\right)}{n} \quad \text{for } i = 1, \cdots, k \text{ and } j = 1, \cdots, k-1$$

and

$$q_{ik} = \begin{cases} \dfrac{p_{ik}-2}{n} = \dfrac{-\left(\varepsilon^{ik}+\bar{\varepsilon}^{ik}\right)}{n} & \text{if } n \text{ is odd} \\[4ex] \dfrac{p_{ik}-1}{n} = \dfrac{-\left(\varepsilon^{ik}+\bar{\varepsilon}^{ik}\right)}{2n} & \text{if } n \text{ is even} \end{cases}$$

It is straightforward to check that

$$P_3 = (3) \qquad\qquad \text{with inverse} \quad P_3^{-1} = \frac{1}{3}(1)$$

$$P_4 = \begin{pmatrix} 2 & 2 \\ 4 & 0 \end{pmatrix} \qquad\qquad \text{with inverse} \quad P_4^{-1} = \frac{1}{4}\begin{pmatrix} 0 & 1 \\ 2 & -1 \end{pmatrix}$$

$$P_5 = \begin{pmatrix} \dfrac{5-\sqrt{5}}{2} & \dfrac{5+\sqrt{5}}{2} \\ \dfrac{5+\sqrt{5}}{2} & \dfrac{5-\sqrt{5}}{2} \end{pmatrix} = \begin{pmatrix} 3-\varphi & 2+\varphi \\ 2+\varphi & 3-\varphi \end{pmatrix} \qquad \text{where } \varphi = \dfrac{1+\sqrt{5}}{2} \text{ is the "Golden number"}$$

$$\text{with inverse} \quad P_5^{-1} = \frac{1}{5}\begin{pmatrix} \dfrac{1-\sqrt{5}}{2} & \dfrac{1+\sqrt{5}}{2} \\ \dfrac{1+\sqrt{5}}{2} & \dfrac{1-\sqrt{5}}{2} \end{pmatrix} = \frac{1}{5}\begin{pmatrix} 1-\varphi & \varphi \\ \varphi & 1-\varphi \end{pmatrix}$$

$$P_6 = \begin{pmatrix} 1 & 3 & 2 \\ 3 & 3 & 0 \\ 4 & 0 & 2 \end{pmatrix} \qquad\qquad \text{with inverse} \quad P_6^{-1} = \frac{1}{6}\begin{pmatrix} -1 & 1 & 1 \\ 1 & 1 & -1 \\ 2 & -2 & 1 \end{pmatrix}$$

$$P_7 = \begin{pmatrix} 2\left(1-\cos\dfrac{2\pi}{7}\right) & 2\left(1-\cos\dfrac{4\pi}{7}\right) & 2\left(1-\cos\dfrac{6\pi}{7}\right) \\ 2\left(1-\cos\dfrac{4\pi}{7}\right) & 2\left(1-\cos\dfrac{6\pi}{7}\right) & 2\left(1-\cos\dfrac{2\pi}{7}\right) \\ 2\left(1-\cos\dfrac{6\pi}{7}\right) & 2\left(1-\cos\dfrac{2\pi}{7}\right) & 2\left(1-\cos\dfrac{4\pi}{7}\right) \end{pmatrix}$$

with inverse

$$P_7^{-1} = -\frac{2}{7}\begin{pmatrix} \cos\dfrac{2\pi}{7} & \cos\dfrac{4\pi}{7} & \cos\dfrac{6\pi}{7} \\ \cos\dfrac{4\pi}{7} & \cos\dfrac{6\pi}{7} & \cos\dfrac{2\pi}{7} \\ \cos\dfrac{6\pi}{7} & \cos\dfrac{2\pi}{7} & \cos\dfrac{4\pi}{7} \end{pmatrix}$$

$$P_8 = \begin{pmatrix} 2-\sqrt{2} & 2 & 2+\sqrt{2} & 2 \\ 2 & 4 & 2 & 0 \\ 2+\sqrt{2} & 2 & 2-\sqrt{2} & 2 \\ 4 & 0 & 4 & 0 \end{pmatrix} \qquad \text{with inverse} \qquad P_8^{-1} = \frac{1}{8}\begin{pmatrix} -\sqrt{2} & 0 & \sqrt{2} & 1 \\ 0 & 2 & 0 & -1 \\ \sqrt{2} & 0 & -\sqrt{2} & 1 \\ 2 & -2 & 2 & -1 \end{pmatrix}$$

$$P_9 = \begin{pmatrix} 2(1-\cos 40°) & 2(1-\cos 80°) & 3 & 2(1-\cos 160°) \\ 2(1-\cos 80°) & 2(1-\cos 160°) & 3 & 2(1-\cos 40°) \\ 3 & 3 & 0 & 3 \\ 2(1-\cos 160°) & 2(1-\cos 40°) & 3 & 2(1-\cos 80°) \end{pmatrix}$$

with inverse

$$P_9^{-1} = \frac{1}{9}\begin{pmatrix} -2\cos 40° & -2\cos 80° & 1 & -2\cos 160° \\ -2\cos 80° & -2\cos 160° & 1 & -2\cos 40° \\ 1 & 1 & -2 & 1 \\ -2\cos 160° & -2\cos 40° & 1 & -2\cos 80° \end{pmatrix}$$

$$P_{10} = \frac{1}{2}\begin{pmatrix} 3-\sqrt{5} & 5-\sqrt{5} & 3+\sqrt{5} & 5+\sqrt{5} & 4 \\ 5-\sqrt{5} & 5+\sqrt{5} & 5+\sqrt{5} & 5-\sqrt{5} & 0 \\ 3+\sqrt{5} & 5+\sqrt{5} & 3-\sqrt{5} & 5-\sqrt{5} & 4 \\ 5+\sqrt{5} & 5-\sqrt{5} & 5-\sqrt{5} & 5+\sqrt{5} & 0 \\ 8 & 0 & 8 & 0 & 4 \end{pmatrix} = \begin{pmatrix} 2-\varphi & 3-\varphi & 1+\varphi & 2+\varphi & 2 \\ 3+\varphi & 2+\varphi & 2+\varphi & 3-\varphi & 0 \\ 1+\varphi & 2+\varphi & 2-\varphi & 3-\varphi & 2 \\ 2+\varphi & 3-\varphi & 3-\varphi & 2+\varphi & 0 \\ 4 & 0 & 4 & 0 & 2 \end{pmatrix}$$

with inverse

$$P_{10}^{-1} = \frac{1}{10} \begin{pmatrix} \dfrac{-1-\sqrt{5}}{2} & \dfrac{1-\sqrt{5}}{2} & \dfrac{-1+\sqrt{5}}{2} & \dfrac{1+\sqrt{5}}{2} & 1 \\[2mm] \dfrac{1-\sqrt{5}}{2} & \dfrac{1+\sqrt{5}}{2} & \dfrac{1+\sqrt{5}}{2} & \dfrac{1-\sqrt{5}}{2} & -1 \\[2mm] \dfrac{-1+\sqrt{5}}{2} & \dfrac{1+\sqrt{5}}{2} & \dfrac{-1-\sqrt{5}}{2} & \dfrac{1-\sqrt{5}}{2} & 1 \\[2mm] \dfrac{1+\sqrt{5}}{2} & \dfrac{1-\sqrt{5}}{2} & \dfrac{1-\sqrt{5}}{2} & \dfrac{1+\sqrt{5}}{2} & -1 \\[2mm] 2 & -2 & 2 & -2 & 1 \end{pmatrix} = \frac{1}{10} \begin{pmatrix} -\varphi & 1-\varphi & -1-\varphi & \varphi & 1 \\ 1-\varphi & \varphi & \varphi & 1-\varphi & -1 \\ -1+\varphi & -\varphi & -\varphi & 1-\varphi & 1 \\ \varphi & 1-\varphi & 1-\varphi & \varphi & -1 \\ 2 & -2 & 2 & -2 & 1 \end{pmatrix}$$

Now we are going to express in a similar manner the index $\|P\|$ in terms of the eigenvalues of the

circulant matrix.

Consider the vector $\vec{v}$ in $\mathbb{R}^n$:

$$\vec{v} = \vec{e}_1 - \frac{1}{n} \sum_{i=1}^{n} \vec{e}_i$$

Given a chord $P$ in $\mathcal{T}^n$, if $M_C(P) = A$, then it is immediate to check that

$$2\|P\|^2 = \|A\vec{v}\|_e^2$$

Now, since $F^* \left( \sum_{i=1}^{n} \vec{e}_i \right) = 0 \quad \Rightarrow \quad F^*(\vec{v}) = \frac{1}{\sqrt{n}}(1,\, 1,\, \cdots,\, 1)^T$

As before $F^* A = D\, A^*$ and

$$2\|P\|^2 = \left\| D\, F^*(\vec{v}) \right\|_e^2 = \frac{1}{n} \sum_{j=1}^{n-1} \left| \lambda_j \right|^2$$

As before since $|\lambda_i| = |\lambda_{n-i}|$, taking $k = \left[\!\!\left[\dfrac{n}{2}\right]\!\!\right]$, and letting $L$ be the $1 \times k$ matrix $L = (\ell_1, \cdots, \ell_k)$

and $\ell_i = 2$, where $i = 1, 2, \cdots, k-1$, and $\ell_k = \begin{cases} 2 & \text{if } n \text{ is odd} \\ 1 & \text{if } n \text{ is even} \end{cases}$

$$\|P\|^2 = \frac{1}{2n} L\left(\lambda^2\right) = \frac{1}{2} L P_n^{-1}\left(\widehat{\sigma}^2\right) , \text{ where } \left(\lambda^2\right), \left(\widehat{\sigma}^2\right), \text{ and } P_n^{-1} \text{ defined as above.}$$

Since $n \displaystyle\sum_{j=1}^{k} q_{ij} = 1$ when $n$ is odd, and $q_{ik} = (-1)^{n-i+1}$ when $n$ is even, we obtain $n L P_n^{-1} = L$.

Hence we obtain the beautiful relation:

$$\boxed{\|P\|^2 \;=\; \frac{1}{2n} L\left(\lambda^2\right) \;=\; \frac{1}{2n} L\left(\widehat{\sigma}^2\right)}$$

Thus we obtain the important conversion formulas:

In $\mathscr{T}^4$, we have

$$\widehat{\sigma_1}^2 = \frac{|\lambda_1|^2 + |\lambda_2|^2}{2} \qquad \text{and} \qquad |\lambda_1|^2 = \widehat{\sigma_2}^2 \qquad \|P\|^2 = \frac{2|\lambda_1|^2 + |\lambda_2|^2}{8}$$

$$\widehat{\sigma_2}^2 = |\lambda_1|^2 \qquad\qquad\qquad |\lambda_2|^2 = 2\widehat{\sigma_1}^2 - \widehat{\sigma_1}^2 \qquad \|P\|^2 = \frac{2\widehat{\sigma_1}^2 + \widehat{\sigma_2}^2}{8}$$

In $\mathscr{T}^6$, we have

$$\widehat{\sigma_1}^2 = \frac{|\lambda_1|^2 + 3|\lambda_2|^2 + 2|\lambda_3|^2}{6} \qquad |\lambda_2|^2 = -\widehat{\sigma_1}^2 + \widehat{\sigma_2}^2 + \widehat{\sigma_3}^2 \qquad \|P\|^2 = \frac{2|\lambda_1|^2 + 2|\lambda_2|^2 + |\lambda_3|^2}{12}$$

$$\widehat{\sigma_2}^2 = \frac{|\lambda_1|^2 + |\lambda_2|^2}{2} \qquad\qquad |\lambda_2|^2 = \widehat{\sigma_1}^2 + \widehat{\sigma_2}^2 - \widehat{\sigma_3}^2 \qquad \|P\|^2 = \frac{2\widehat{\sigma_1}^2 + 2\widehat{\sigma_2}^2 + \widehat{\sigma_3}^2}{12}$$

$$\widehat{\sigma_3}^2 = \frac{2|\lambda_1|^2 + |\lambda_3|^2}{3} \qquad\qquad |\lambda_3|^2 = 2\widehat{\sigma_1}^2 - 2\widehat{\sigma_2}^2 + \widehat{\sigma_3}^2$$

In $\mathcal{T}^8$, we have

$$\widehat{\sigma_1}^2 = \frac{(2-\sqrt{2})|\lambda_1|^2 + 2|\lambda_2|^2 + (2+\sqrt{2})|\lambda_3|^2 + 2|\lambda_4|^2}{8}$$

$$|\lambda_1|^2 = -\sqrt{2}\,\widehat{\sigma_1}^2 + \sqrt{2}\,\widehat{\sigma_3}^2 + \widehat{\sigma_4}^2$$

$$\widehat{\sigma_2}^2 = \frac{|\lambda_1|^2 + 2|\lambda_2|^2 + |\lambda_3|^2}{4}$$

$$|\lambda_2|^2 = 2\widehat{\sigma_2}^2 - \widehat{\sigma_4}^2$$

$$\widehat{\sigma_3}^2 = \frac{(2+\sqrt{2})|\lambda_1|^2 + 2|\lambda_2|^2 + (2-\sqrt{2})|\lambda_3|^2 + 2|\lambda_4|^2}{8}$$

$$|\lambda_3|^2 = \sqrt{2}\,\widehat{\sigma_1}^2 - \sqrt{2}\,\widehat{\sigma_3}^2 + \widehat{\sigma_4}^2$$

$$\widehat{\sigma_4}^2 = \frac{|\lambda_1|^2 + |\lambda_3|^2}{2}$$

$$|\lambda_4|^2 = 2\widehat{\sigma_1}^2 - 2\widehat{\sigma_2}^2 + 2\widehat{\sigma_3}^2 - \widehat{\sigma_4}^2$$

$$\|P\|^2 = \frac{2|\lambda_1|^2 + 2|\lambda_2|^2 + 2|\lambda_3|^2 + |\lambda_4|^2}{16} = \frac{2\widehat{\sigma_1}^2 + 2\widehat{\sigma_2}^2 + 2\widehat{\sigma_3}^2 + \widehat{\sigma_4}^2}{16}$$

## Classification of Degenerated Chords

For $n \geq 2$, we have the following definitions:

A Chord $P$ in $\mathcal{T}^n$ is degenerated $\quad \Leftrightarrow \quad \det(M_C(P)) = 0$, or $\det(M_H(P)) = 0$

A Chord $P$ in $\mathcal{T}^n$ $(n|12)$ is barycentric $\quad \Leftrightarrow \quad \operatorname{rank}(M_C(P)) = \operatorname{rank}(M_H(P)) = 1$

A Chord $P$ in $\mathcal{T}^n$ is polar $\quad \Leftrightarrow \quad P$ is degenerated,

and $\operatorname{rank}(M_C(P)) = \operatorname{rank}(M_H(P)) = 2$

A Chord $P$ in $\mathcal{T}^n$ is $\mathbb{S}^k$-equatorial $\quad \Leftrightarrow \quad P$ is degenerated,

and $\operatorname{rank}(M_C(P)) = \operatorname{rank}(M_H(P)) = k+2$

We derive the following conclusions:

**1)** If $n$ is prime and $n > 3$, $\mathcal{T}^n$ does not contain degenerated chords.

Note that the only degenerated chords in $\mathcal{T}^2$ and $\mathcal{T}^3$ are their barycenters $(6, 6)$ and $(4, 4, 4)$.

From linear algebra, if $\text{rank}\left(M_C(P)\right) = n - k$, then $M_C(P)$ has exactly $k$ eigenvalues that are

zero. If $\lambda_\ell = \sum_{k=1}^{n} x_k \varepsilon^{(k-1)\ell} = 0$ , $\ell = 0, 1, 2, \cdots, n-1$ , since $n$ is prime and $\varepsilon^\ell$ is a primitive root

of unity, the multiples of $\varepsilon^\ell$ generate all of the different $n^{\text{th}}$-roots of unity. But since

$\sum_{k=1}^{n} \varepsilon^{(k-1)\ell} = 0$, and the $x_k$ are in , the $x_k$ must be equal and therefore $P$ is a barycenter. But if $n$

is prime and $n > 3$, then there are no barycentric chords.

**2)** If a chord $P$ in $\mathcal{T}^n$ is $\mathbb{S}^{n-3}$-equatorial, then

$$x_1 - x_2 + \cdots + x_{n-1} - x_n = 0 \qquad \text{or equivalently} \qquad \sum_{k=1}^{n/2} x_{2k-1} = 6$$

In fact if $P$ is $\mathbb{S}^{n-3}$-equatorial, the $\text{rank}\left(M_C(P)\right) = n - 1$, and there is only one zero eigenvalue.

But since $\lambda_0 = 12$ and $\lambda_i = \overline{\lambda_{n-i}}$ , the only possible zero eigenvalue is

$\lambda_{n/2} = x_1 - x_2 + \cdots + x_{n-1} - x_n$. Notice that $\mathbb{S}^{n-3}$-equatorial chords exist in $\mathcal{T}^n$ only when $n$ is

even. For example a non-barycentric chord $P$ in $\mathcal{T}^4$ is $\mathbb{S}^1$-equatorial $\qquad \Leftrightarrow$

$x_1 - x_2 + x_3 - x_4 = 0$, or equivalently $x_1 + x_3 = 6$.

**3)** More generally, a non-barycentric chord $P$ in $\mathcal{T}^n$, where $n$ is even is $\mathbb{S}^{n-2k-1}$-equatorial for

some $k = 1, 2, \cdots, \dfrac{n}{2} - 1 \qquad \Leftrightarrow \qquad x_1 - x_2 + \cdots + x_{n-1} - x_n = 0$

Now we are going to classify the non-barycentric degenerated chords in terms of the orbital

meshes $\widehat{\sigma_k}$ and the eigenvalues of the cyclic permutation matrices $\lambda_k$.

In $\mathcal{T}^4$, we have

A chord is polar $\quad \Leftrightarrow \quad \lambda_1 = \lambda_3 = 0$, and $\lambda_2 \neq 0 \Leftrightarrow \widehat{\sigma_2} = 0$

A chord is $\mathbb{S}^1$-equatorial $\quad \Leftrightarrow \quad \lambda_2 = 0 \Leftrightarrow \widehat{\sigma_2} = \sqrt{2}\,\widehat{\sigma_1} \qquad$ or $\left(\widehat{\sigma_2} = 2\|P\|\right)$

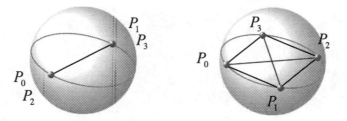

**Figure 4.2** The vertices for the degenerated chords are shown with $P_k = \sigma^k(P)$.

In $\mathcal{T}^6$, we have

A chord is polar $\quad \Leftrightarrow \quad \lambda_1 = \lambda_2 = \lambda_4 = \lambda_5 = 0$, and $\lambda_3 \neq 0 \Leftrightarrow \widehat{\sigma_2} = 0$

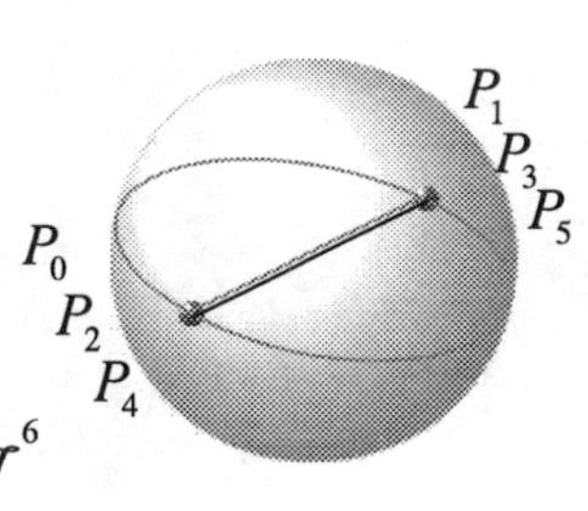

**Figure 4.3**
A polar chord in in $\mathcal{T}^6$

A chord is $\mathbb{S}^1$-equatorial $\qquad \Leftrightarrow \quad \lambda_3 = 0$ and $\begin{cases} \lambda_1 = 0 \\ \text{or} \\ \lambda_2 = 0 \end{cases} \Leftrightarrow \begin{cases} \widehat{\sigma_3} = 0 & \text{triangle} \\ \text{or} \\ \widehat{\sigma_3} = 2\,\widehat{\sigma_1} & \text{hexagon} \end{cases}$

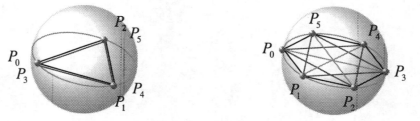

**Figure 4.4** $\mathbb{S}^1$-equatorial chords in $\mathcal{T}^6$.

53

A chord is $\mathbb{S}^2$-equatorial $\qquad \Leftrightarrow \qquad \begin{cases} \lambda_2 = 0 \\ \text{and} \\ \lambda_1 \cdot \lambda_3 \neq 0 \end{cases} \qquad \Leftrightarrow \qquad \widehat{\sigma_3}^2 = \widehat{\sigma_1}^2 + \widehat{\sigma_2}^2$

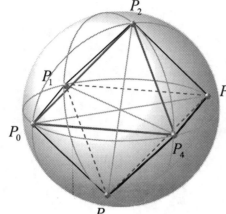

Figure 4.5 $\mathbb{S}^2$-equatorial chords in $\mathscr{T}^6$

A chord is $\mathbb{S}^3$-equatorial $\Leftrightarrow \begin{cases} \lambda_3 = 0 \\ \text{and} \\ \lambda_1 \cdot \lambda_2 \neq 0 \end{cases} \Leftrightarrow \begin{cases} \widehat{\sigma_3} \neq 0, \text{ and } \widehat{\sigma_3} \neq 2\widehat{\sigma_1} \\ \text{and} \\ \widehat{\sigma_3}^2 = 2\widehat{\sigma_2}^2 - 2\widehat{\sigma_1}^2 \\ \text{or } \left( \widehat{\sigma_2} = \sqrt{3}\|P\| \right) \end{cases}$

In $\mathscr{T}^8$, we have

A chord is polar $\qquad \Leftrightarrow \qquad \lambda_1 = \lambda_2 = \lambda_3 = \lambda_5 = \lambda_6 = \lambda_7 = 0$, and $\lambda_4 \neq 0 \Leftrightarrow \widehat{\sigma_2} = 0$

Figure 4.6
A polar chord in in $\mathscr{T}^8$.

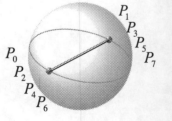

A chord is $\mathbb{S}^1$-equatorial $\Leftrightarrow \begin{cases} \lambda_4 = 0 \\ \lambda_1 = \lambda_3 = 0 \\ \lambda_2 \neq 0 \end{cases} \Leftrightarrow \begin{cases} \widehat{\sigma_4} = 0 \\ \text{and} \\ \widehat{\sigma_2} = \sqrt{2}\widehat{\sigma_1} \end{cases}$

Figure 4.7
$\mathbb{S}^1$-equatorial chord in in $\mathscr{T}^8$.

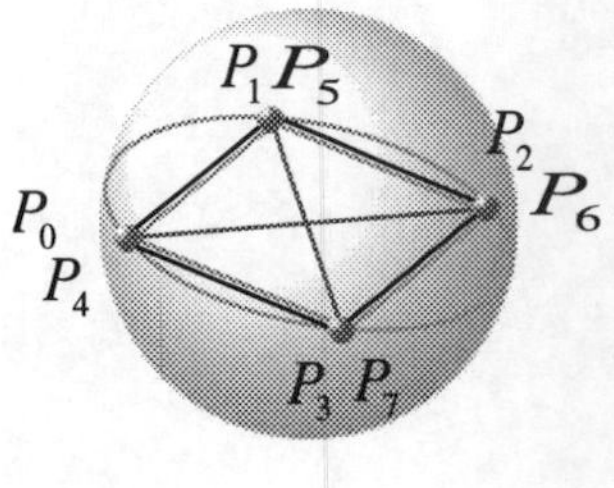

A chord is $\mathbb{S}^2$-equatorial $\Leftrightarrow$ $\begin{cases} \lambda_1 = \lambda_3 = 0 \\ \lambda_2 \cdot \lambda_4 \neq 0 \end{cases}$ $\Leftrightarrow$ $\begin{cases} \widehat{\sigma_4} = 0 \\ \text{and} \\ \widehat{\sigma_2} \neq \sqrt{2}\,\widehat{\sigma_1} \end{cases}$

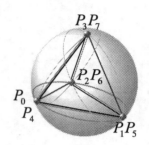

**Figure 4.8**
$\mathbb{S}^2$-equatorial chord in $\mathscr{T}^8$

A chord is $\mathbb{S}^3$-equatorial $\Leftrightarrow$ $\begin{cases} \lambda_2 = 0 \\ \lambda_4 = 0 \\ \lambda_1 \cdot \lambda_3 \neq 0 \end{cases}$ $\Leftrightarrow$ $\begin{cases} \widehat{\sigma_4} \neq 0 \\ \widehat{\sigma_1}^2 - 2\widehat{\sigma_2}^2 + \widehat{\sigma_3}^2 = 0 \end{cases}$

A chord is $\mathbb{S}^4$-equatorial $\Leftrightarrow$ $\begin{cases} \lambda_2 = 0 \\ \lambda_1 \cdot \lambda_3 \cdot \lambda_4 \neq 0 \end{cases}$ $\Leftrightarrow$ $\begin{cases} \widehat{\sigma_4} = \sqrt{2}\,\widehat{\sigma_2} \\ \text{and} \\ \widehat{\sigma_1}^2 - 2\widehat{\sigma_2}^2 + \widehat{\sigma_3}^2 \neq 0 \end{cases}$

A chord is $\mathbb{S}^5$-equatorial $\Leftrightarrow$ $\begin{cases} \lambda_4 = 0 \\ \lambda_1 \cdot \lambda_2 \cdot \lambda_3 \neq 0 \end{cases}$ $\Leftrightarrow$ $\begin{cases} \widehat{\sigma_4} \neq \sqrt{2}\,\widehat{\sigma_2} \\ \widehat{\sigma_4} \neq 0 \\ \widehat{\sigma_4}^2 = 2\widehat{\sigma_1}^2 - 2\widehat{\sigma_2}^2 + 2\widehat{\sigma_3}^2 \end{cases}$ $\text{or}\left( \|P\| = \dfrac{\sqrt{\widehat{\sigma_1}^2 + \widehat{\sigma_3}^2}}{2} \right)$

In $\mathscr{T}^9$, we have

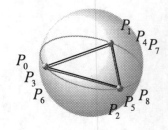

**Figure 4.9**

$\mathbb{S}^1$-equatorial chord in in $\mathscr{T}^9$

A chord is $\mathbb{S}^1$-equatorial $\Leftrightarrow$
$\begin{cases} \lambda_1 = 0 \\ \lambda_2 = 0 \\ \lambda_4 = 0 \\ \lambda_3 \neq 0 \end{cases}$
$\Leftrightarrow$ $\widehat{\sigma_3} = 0$

A chord is $\mathbb{S}^5$-equatorial $\Leftrightarrow$
$\begin{cases} \lambda_3 = 0 \\ \lambda_1 \cdot \lambda_2 \cdot \lambda_4 \neq 0 \end{cases}$
$\Leftrightarrow$ $\widehat{\sigma_3}^2 = \dfrac{\widehat{\sigma_1}^2 + \widehat{\sigma_2}^2 + \widehat{\sigma_4}^2}{2}$ or $\widehat{\sigma_3} = \sqrt{3}\|P\|$

In $\mathscr{T}^{10}$, we have

A chord is $\mathbb{S}^3$-equatorial $\Leftrightarrow$
$\begin{cases} \lambda_1 = 0 \\ \lambda_3 = 0 \\ \lambda_5 = 0 \\ \lambda_2 \cdot \lambda_4 \neq 0 \end{cases}$
$\Leftrightarrow$ $\widehat{\sigma_5} = 0$

A chord is $\mathbb{S}^7$-equatorial $\Leftrightarrow$
$\begin{cases} \lambda_5 = 0 \\ \lambda_1 \cdot \lambda_2 \cdot \lambda_3 \cdot \lambda_4 \neq 0 \end{cases}$
$\Leftrightarrow$ $\widehat{\sigma_5}^2 = 2\left( \widehat{\sigma_4}^2 - \widehat{\sigma_3}^2 + \widehat{\sigma_2}^2 - \widehat{\sigma_1}^2 \right)$

or $\left( \|P\| = \dfrac{\sqrt{\widehat{\sigma_2}^2 + \widehat{\sigma_4}^2}}{5} \right)$

## 5. Chord Complementation

Consider a chord $P = (x_1, x_2, \cdots, x_n)$ in $\mathcal{T}^n$ and assume this chord has precisely $x_{i_1}, x_{i_2}, \cdots, x_{i_k}$ intervals greater than one. If we assume that $P$ is given in prime form, then $x_n > 1$ and thus we can consider $x_{i_k} = x_n$. We can rewrite $P$ as:

$$P = \left( \underbrace{1, 1, \cdots, 1}_{i_1 - 1}, x_{i_1}, \underbrace{1, 1, \cdots, 1}_{i_2 - i_1 - 1}, x_{i_2}, \cdots, x_{i_{k-1}}, \underbrace{1, 1, \cdots, 1}_{i_k - i_{k-1} - 1}, x_{i_k} \right)$$

Define the complementation map $\mathcal{C} : \mathcal{T}^n \to \mathcal{T}^{12-n}$ as the map that takes $P$ into $P^C$, where

$$P^C = \left( \underbrace{1, 1, \cdots, 1}_{x_{i_1} - 2}, 2 + i_2 - i_1 - 1, \underbrace{1, 1, \cdots, 1}_{x_{i_2} - 2}, 2 + i_3 - i_2 - 1, \cdots, \underbrace{1, 1, \cdots, 1}_{x_{i_{k-1}} - 2}, 2 + i_k - i_{k-1} - 1, \underbrace{1, 1, \cdots, 1}_{x_{i_k} - 2}, 2 + i_1 - 1 \right)$$

At first glance this definition appears to be very complicated. However the notation makes this, otherwise beautiful map, more complicated than it really is. The complementation $\mathcal{C}$ breaks every interval $x_i$ different from 1, into $x_i - 2$ consecutive 1's and then it is immediately followed by the interval $2 + p$, where $p$ is the number of consecutive 1's after $x_i$ in the original chord $P$.

**Example 5.1**

$P = (3, 3, 1, 1, 4) \quad \longrightarrow \quad P^C = ( \underset{3-2}{1} , \underset{2+0}{2} , \underset{3-2}{1} , \underset{2+2}{4} , \underbrace{1, 1}_{4-2} , \underset{2+0}{2} )$

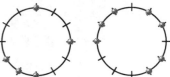

**Figure 5.1a**
Clock representation of $P$ and $P^C$

$P = (2, 1, 3, 1, 1, 4) \quad \longrightarrow \quad P^C = ( \underset{\substack{2-2 \\ 1's}}{3} , \underset{2+1}{1} , \underset{3-2}{4} , \underset{2+2}{1} , \underbrace{1, 1}_{4-2} , \underset{2+0}{2} )$

**Figure 5.1b**
Clock representation of $P$ and $P^C$

57

The map $\mathcal{C}$ is well-defined:

Effectively, since $P$ is in $\mathcal{T}^n$,

$$\sum_{j=1}^{k} x_{i_j} + i_1 - 1 + i_2 - i_1 - 1 + \cdots + i_k - i_{k-1} - 1 = 12$$

After cancellation we obtain

$$\sum_{j=1}^{k} x_{i_j} + n - k = 12 \qquad (5.1)$$

Now if we add the terms in $P^C$ we have

$$\sum_{j=1}^{k} x_{i_j} - 2 + \sum_{j=1}^{k}\left(2 + i_{j+1} - i_j - 1\right) + 2 + i_1 - 1 \; = \; \left(\sum_{j=1}^{k} x_{i_j}\right) - 2k + k + i_k$$

Since $i_k = n$, this sum equals 12. (5.1)

On the other hand, the number of intervals in $P^C$ is

$$\sum_{j=1}^{k} x_{i_j} - 2k + k = 12 - n$$

by (5.1). So $P^C$ is a legitimate member of $\mathcal{T}^{12-n}$ and the map $\mathcal{C}$ is well-defined.

The complementation name in this map is properly justified with the example below:

Consider $P = (1, 1, 3, 3, 4)$ in $\mathcal{T}^5$, which written in pitch-class notation would appears as $[0, 1, 2, 5, 8]$. Now $P^C = (1, 2, 1, 2, 1, 1, 4)$ or in pitch-class notation $[0, 1, 3, 4, 6, 7, 8]$ which is the pitch-class complement of $[0, 1, 2, 5, 8]$. Notice that the transposition:

$T^9[0, 1, 2, 5, 8]=[9, A, B, 2, 5]$.


Now, if we observe the chords $P$ and $P^C$ in the complementation definition we can derive easily:

$$\mathcal{C}\left(\sigma^\ell(P)\right) = \mathcal{C}(P) \qquad \text{, for } \ell = 1, 2, \cdots, i_1 - 1$$

and

$$\mathcal{C}\left(\sigma^{i_1}(P)\right) = \sigma^{x_{i_1}-1}\left(\mathcal{C}(P)\right)$$

The above identities suggest the extension of the complementation operation to the orbits $\overline{P}$ in $\mathcal{T}^n / \Sigma^n$

$$\mathcal{C}\left(\overline{P}\right) = \overline{\mathcal{C}(P)}$$

The map $\mathcal{C}$ is idempotent, i.e. $\mathcal{C}^2\left(\overline{P}\right)=\overline{P}$. In the notation above

$$P^C = \left( \underbrace{1, 1, \cdots, 1}_{x_{i_1}-2}, 2+i_2-i_1-1, \underbrace{1, 1, \cdots, 1}_{x_{i_2}-2}, \cdots, \underbrace{1, 1, \cdots, 1}_{x_{i_k}-2}, 2+i_1-1 \right)$$

and now,

$$\mathcal{C}\left(P^C\right) = \left( \underbrace{1, 1, \cdots, 1}_{i_2-i_1-1}, 2+x_{i_2}-2, \underbrace{1, 1, \cdots, 1}_{i_3-i_2-1}, \cdots, 2+x_n-2, \underbrace{1, 1, \cdots, 1}_{i_1-1} \right)$$

which equals

$$\sigma^{n-i_1+1}(P)$$

and thus $\mathcal{C}^2\left(\overline{P}\right)=\overline{P}$.

Consider $P$ in $\mathcal{T}^n$ as above, and construct its inversion:

$$I(P) = \left( x_{i_k}, \underbrace{1, 1, \cdots, 1}_{i_k-i_{k-1}-1}, \cdots, x_{i_2}, \underbrace{1, 1, \cdots, 1}_{i_2-i_1-1}, x_{i_1}, \underbrace{1, 1, \cdots, 1}_{i_1-1} \right)$$

and

$$\left(I(P)\right)^C = \left( \underbrace{1, 1, \cdots, 1}_{x_{i_k}-2}, 2+i_k-i_{k-1}-1, \cdots, \underbrace{1, 1, \cdots, 1}_{x_{i_2}-1}, 2+i_2-i_1-1, \underbrace{1, 1, \cdots, 1}_{x_{i_1}-2}, 2+i_1-1 \right)$$

Now,

$$P^C = \left( \underbrace{1, 1, \cdots, 1}_{x_{i_k}-2}, 2+i_2-i_1-1, \cdots, 2+i_k-i_{k-1}-1, \underbrace{1, 1, \cdots, 1}_{x_{i_k}-2}, 2+i_1-1 \right)$$

and thus,

$$I(P^C) = \left( 2+i_1-1, \underbrace{1, 1, \cdots, 1}_{x_{i_k}-2}, 2+i_k-i_{k-1}-1, \cdots, 2+i_2-i_1-1, \underbrace{1, 1, \cdots, 1}_{x_{i_k}-2} \right)$$

So we have

$$\sigma \circ I\left(P^C\right) = \left(I(P)\right)^C$$

Thus if $P$ and $Q$ are co-$I$-symmetric in $\mathcal{T}^n/\Sigma^n$, i.e. $I\left(\overline{P}\right) = \overline{Q}$, then their complements would be co-$I$-symmetric in $\mathcal{T}^{12-n}/\Sigma^{12-n}$, i.e. $\mathcal{C}\left(\overline{Q}\right) = I\left(\mathcal{C}\left(\overline{P}\right)\right)$.

Moreover if $P$ is $I$-symmetric in $\mathcal{T}^n/\Sigma^n$, then $P^C$ is $I$-symmetric in $\mathcal{T}^{12-n}/\Sigma^{12-n}$.

**Complementation and degeneracy**

Now since the complementation is properly introduced, it will be interesting to analyze the complementation of barycentric and polar chords.

Let $B$ be a barycetric chord in $\mathcal{T}^n$, where $n \leq 6$ and $n \mid 12$. (Note that the chromatic scale is barycentric in $\mathcal{T}^{12}$ with an empty complement.)

$$B = \left(\frac{12}{n}, \cdots, \frac{12}{n}\right) \quad \text{and} \quad B^C = \left(\underbrace{1, 1, \cdots, 1}_{\frac{12}{n}-2}, 2, \cdots, \underbrace{1, 1, \cdots, 1}_{\frac{12}{n}-2}, 2\right)$$

Observe that $\operatorname{rank} M_H\left(B^C\right) = \dfrac{12}{n} - 1$

Thus,

If $n = 2$, then $B^C$ is $S^3$-equatorial in $\mathcal{T}^{10}$

If $n = 3$, then $B^C$ is $S^1$-equatorial in $\mathcal{T}^9$

If $n = 4$, then $B^C$ is polar in $\mathcal{T}^8$

If $n = 6$, then $B^C$ is barycentric in $\mathcal{T}^6$

If $P = \left(x_1, x_2, \cdots, x_n\right)$ is polar in $\mathcal{T}^n$, where $n = 4, 6, 8$

$$\widehat{\sigma_2}(P) = 0 \quad \text{and} \quad x_i = x_{i+2}$$

Assume without loss of generality that $x_1 \neq 1$.

If $x_2 = 1$, then

$$P^C = \left( \underbrace{1, 1, \cdots, 1}_{x_1 - 2}, 3, \underbrace{1, 1, \cdots, 1}_{x_3 - 2 = x_1 - 2}, 3, \cdots \right)$$

and rank $M_H(P^C) = x_1 - 2 + 1 = x_1 - 1$

If $x_2 \neq 1$, then

$$P^C = \left( \underbrace{1, 1, \cdots, 1}_{x_1 - 2}, 2, \underbrace{1, 1, \cdots, 1}_{x_2 - 2}, 2, \underbrace{1, 1, \cdots, 1}_{x_3 - 2 = x_1 - 2}, 2, \cdots \right)$$

and rank $M_H(P^C) = x_1 - 2 + x_2 - 2 + 2 = x_1 + x_2 - 2$

In both cases we obtain:

$$\boxed{\text{rank } M_H(P^C) = x_1 + x_2 - 2}$$

If $n = 4$, then $x_1 + x_2 = 6$, and $P^C$ is $\mathbb{S}^2$-equatorial.

If $n = 6$, then $x_1 + x_2 = 4$, and $P^C$ is polar.

If $n = 8$, then $x_1 + x_2 = 3$, and $P^C$ is barycentric.

## Complementation with Enrichments and Reductions

We are going to state four very important properties that relate complementation $\mathcal{C}$ with the compositions of inversion $I$, enrichments $f_k$, and reductions $\pi_k$. Note that the subscript $k$ denotes the enumeration of enrichments or reductions operated on chord $P$.

Let $P* = (1,2,2,1,6) = 5\text{-}Z12$ with complement $P*^C = (1,1,1,1,3,2,3) = 7\text{-}Z12$.

---

Given a chord $P$ in $\mathcal{T}^n$, where $n < 6$, and $P$ is not in the orbit $\overline{P*}$, we can find a representative $P^C$ of $\overline{P}^C$ and $12 - 2n$ enrichments such that

$$P^C = f_k \circ f_{k-1} \circ \cdots \circ f_1 \circ I(P) \quad , \text{or } I\left(P^C\right) = f_k \circ f_{k-1} \circ \cdots \circ f_1(P) \qquad , \text{where } k = 12 - 2n \quad (5.2)$$

---

**Example 5.2**

$$P = (1,\ 2,\ 4,\ 5) \in \mathcal{T}^4, \text{ with } P^C = (2,\ 1,\ 1,\ 2,\ 1,\ 1,\ 1,\ 3) \in \mathcal{T}^8$$

In order to construct the complement, first take the inversion of $P$, then perform the following compositions of enrichments:

$$
\begin{array}{c}
I(P) = \quad (5,\ 4,\ 2,\ 1) \\
f_{\#1} \downarrow \\
(\overbrace{3,\ 2}^{},\ 4,\ 2,\ 1) \\
f_{\#2} \downarrow \\
(3,\ 2,\ \overbrace{2,\ 2}^{},\ 2,\ 1) \\
f_{\#3} \downarrow \\
(3,\ 2,\ \overbrace{1,\ 1}^{},\ 2,\ 2,\ 1) \\
f_{\#4} \downarrow \\
(3,\ 2,\ 1,\ 1,\ 2,\ \overbrace{1,\ 1}^{},\ 1) = \sigma^7\left(P^C\right)
\end{array}
$$

Notice that in this case the initial inversion is essential. Without this inversion it would be impossible to arrive at $P^C$.

## Proof of complementation result (5.2)

We are going to break down the proof according to the number of intervals different from 1 in the chord.

*CASE 1.* Consider a chord $P$ in $\mathcal{T}^n$, $\overline{n}$ 6 with only one interval $x$ different from 1.

Let $p$ be the number of ones in $P$. If $p = 0$, then $x = 12$ and $10 = 12 - 2(1)$ enrichments will map $I(P) = (x)$ into $(\underbrace{1,1,...,1}_{10},2) \in \mathcal{T}^{11}$.

If $1 \le p \le 4$, then $P = (\underbrace{1,1,...,1}_{p},x)$ with $P^C = (\underbrace{1,1,...,1}_{x-2},2+p)$.

Since $x \ge 6$:  $\quad I(P) = (x,\overbrace{1,1,...,1}^{p})$

$$\downarrow f_1$$

$$(\overbrace{x-2-p,2+p,\underbrace{1,1,...,1}_{p}})$$

$x$-2-$p$ will reduce to ones after $x$-2-$p$-1 enrichments. Thus the total number of enrichments is $x$-2-$p$ = 12-$p$, since $x$ =12-$p$ and $p = n$-1.

*CASE 2.* Now we have two intervals $x_1$ and $x_2$ different from one and $p$ ones in our chord $P$.

If $p = 0$ then $P = (x_1,x_2)$ with complement $P^C = (\overbrace{1,1,...,1}^{x_1-2},2,\overbrace{1,1,...,1}^{x_2-2},2)$.
and so we break down $I(P) = (x_2,x_1)$

$$f_1 \swarrow \qquad \searrow f_2$$

$$(\overbrace{x_2-2,2},\overbrace{x_1-2,2})$$

and $x_2 - 2$ and $x_1 - 2$ will become ones after $x_1 - 3$ and $x_2 - 3$ enrichments respectively.

The total number of enrichments is $2 + x_1 - 3 + x_2 - 3 = 12$-2$n$ .

Now if $1 \le p \le 3$ consider two subcases:

A.  Consecutive ones.

$$P = (\overbrace{1,1,...,1}^{p}, x_1, x_2) \text{ with } P^C = (\overbrace{1,1,...,1}^{x_1-2}, 2, \overbrace{1,1,...,1}^{x_2-2}, 2+p)$$

Assume without any loss of generality that $x_2 \ge 2+p$

Then $I(P) = (x_2, x_1, \overbrace{1,1,...,1}^{p})$ with enrichments

$$f_1 \swarrow \qquad \searrow f_2$$

$$(\overbrace{x_2 - 2 - p, 2 + p}, \overbrace{x_1 - 2, 2}, \overbrace{1,1,...,1}^{p}) \quad \text{and now converting } x_2 - 2 - p \text{ and } x_1 - 2 \text{ to ones we will}$$

have a total of $2 + x_2 - 3 - p + x_1 - 3 = 12\text{-}2n$ enrichments.

B.  Non-consecutive ones.

$$P = (1, x_1, \overbrace{1,...,1}^{p}, x_2) \text{ with } P^C = (\overbrace{1,1,...,1}^{x_1-2}, 2+p, \overbrace{1,1,...,1}^{x_2-2}, 3)$$

The case $p = 0$ was proven in part A above.

If $1 \le p \le 2$ then $I(P) = (x_2, \overbrace{1,...,1}^{p}, x_1, 1)$ and assuming $x_2 \ge 2+p$ we enrich as follows:

$$f_1 \swarrow \qquad \searrow f_2$$

$$(\overbrace{2+p, x_2 - 2 - p}, \overbrace{1,...,1}^{p}, 3, x_1 - 3, 1) \quad \text{if} \qquad \boxed{x_1 \ge 3}$$

Converting $x_2 - 2 - p$ and $x_1 - 3$ to ones we will get a total of $2 + x_2 - 2 - p - 1 + x_1 - 3 - 1 = 12\text{-}2n$ enrichments, since $p = n - 3$.

If $\boxed{x_1 = 2}$ above we will have:

$$P = (1, 2, \overbrace{1,...,1}^{p}, x_2) \text{ with } P^C = (2 + p, \overbrace{1,1,...,1}^{x_2-2}, 3)$$

Now we have:

$I(P) = (x_2, \overbrace{1,...,1}^{p}, 2, 1)$  with enrichments:

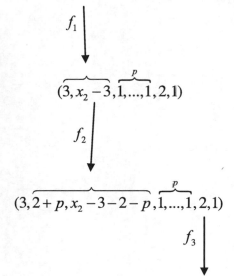

$f_1$

$(3, x_2 - 3, \overbrace{1,...,1}^{p}, 2, 1)$

$f_2$

$(3, \overbrace{2 + p, x_2 - 3 - 2 - p}, \overbrace{1,...,1}^{p}, 2, 1)$

$f_3$

$(3, 2 + p, x_2 - 3 - 2 - p, \overbrace{1,...,1}^{p}, \overbrace{1,1}, 1)$ and converting $x_2 - 3 - 2 - p$ to ones, we arrive at the complementary chord, after $3 + x_2 - 5 - p - 1 = 12 - 2n$ enrichments. ($p = n - 3$).

*CASE 3.* We treat now chords with three intervals $x_1, x_2, x_3$ different from one.

Let $p$, as before, be the number of ones in $P$.

If $p = 0$ then $P = (x_1, x_2, x_3)$ with $P^C = (\overbrace{1, 1..., 1}^{x_1 - 2}, 2, \overbrace{1, 1, ..., 1}^{x_2 - 2}, 2, \overbrace{1, 1, ..., 1}^{x_3 - 2}, 2)$.

Assume $x_3 \geq x_2$, then we can enrich $I(P) = (x_3, x_2, x_1)$ in the following manner:

$f_1$

$(\overbrace{x_2 - 2, x_3 - x_2 + 2}, x_2, x_1)$

$f_2$

$(x_2 - 2, \overbrace{2, x_3 - x_2}, x_2, x_1)$

$f_3$     $f_4$

$(x_2 - 2, 2, x_3 - x_2, \overbrace{x_2 - 2, 2}, \overbrace{x_1 - 2, 2})$    and converting $x_2 - 2$ and

$x_1 - 2$ to ones we obtain a representative of $\overline{P}^C$ after

$4 + x_2 - 3 + x_3 - x_2 - 1 + x_2 - 3 + x_1 - 3 = x_1 + x_2 + x_3 - 6 = 12 - 2n$ enrichments.

Now if $p \neq 0, 1 \le p \le 2$ and then, as before, we can consider two subcases:

A. Consecutive ones.

$$P = (\overbrace{1,...,1}^{p}, x_1, x_2, x_3) \text{ with } P^C = (1,1,...,1,2,\overbrace{1,1,...,1}^{x_1-2},2,\overbrace{1,1,...,1}^{x_2-2},2,\overbrace{1,1,...,1}^{x_3-2},2+p).$$

Assume firstly that, then we construct the following sequence of enrichments:

$$I(P) = (x_3, x_2, x_1, \overbrace{1,...,1}^{p})$$

$$f_1 \downarrow$$

$$(\overbrace{x_3 - 2 - p, 2 + p}, x_2, x_1, \overbrace{1,...,1}^{p})$$

$$f_2 \downarrow$$

$$(\overbrace{x_3 - 2 - p, 2 + p, \overbrace{x_1 - 2, x_2 - x_1 + 2}}, x_1, \overbrace{1,...,1}^{p}) \qquad \text{if } \boxed{x_2 \ge x_1}$$

$$f_3 \downarrow \qquad \diagdown f_4$$

$$(x_3 - 2 - p, 2 + p, x_1 - 2, \overbrace{2, x_2 - x_1}, \overbrace{x_1 - 2, 2}, \overbrace{1,...,1}^{p})$$

Converting $x_3 - 2 - p, x_1 - 2, x_2 - x_1, x_1 - 2$ to ones we arrive at a representative of $\overline{P}^C$ after $4 + x_1 + x_2 + x_3 - 6 - p - 4 = 12 - 2p - 6 = 12 - 2n$ enrichments.

Now if $\boxed{x_1 > x_2}$, we consider first the case $x_2 = 2$

Thus now our chord $P = (\overbrace{1,...,1}^{p}, x_1, 2, x_3)$ will have

complement: $P^C = (1,1,...,1,2,2,\overbrace{1,1,...,1}^{x_1-2},2,2,\overbrace{1,1,...,1}^{x_3-2},2+p)$

Assume firstly that $x_1 \ge 2 + p$.

$$I(P) = (x_3, 2, x_1, \overbrace{1, \ldots, 1}^{p})$$

$f_1 \downarrow$

$$(\overbrace{2+p, x_1-2-p}, \overbrace{1, \ldots, 1}^{p}, x_3, 2)$$

$f_2 \downarrow \qquad\qquad\qquad \text{if } x_3 \geq 4$

$$(2+p, x_1-2-p, \overbrace{1, \ldots, 1}^{p}, \overbrace{4, x_3-4}, 2)$$

$f_3 \downarrow$

$$(2+p, x_1-2-p, \overbrace{1, \ldots, 1}^{p}, \overbrace{2, 2}, x_3-4, 2) \quad \text{and converting } x_1-2-p, x_3-4 \text{ and } 2 \text{ to ones we get}$$

to a representative of $\bar{P}^C$ after $3 + 1 + x_1 - 2 - p - 1 + x_3 - 4 - 1 = 12 - 2n$ enrichments.

It is important to observe at this point that the inversion properties of the enrichments (3.3) guarantee the following immediate statement: A chord $P$ satisfies (5.2) if and only if $I(P)$ satisfies the same.

If $x_3 < 4$ above, the only situation where the procedure fails will be with the chord $(1,6,2,3)$, but we should notice that $(1,3,2,6)$, a representative of the inversion, satisfies the condition above and the inversion argument takes care of this case.

Now if we assume $x_1 < 2 + p$ then $x_1 = 3$ and in the present conditions we have only two chords left: $(1,3,2,6)$ and $(1,1,3,2,5)$. We have dealt with the former one already. Notice that the two enrichments: $f_4^2 \circ f_1^4 (5,2,3,1,1) = (4,1,2,2,1,1,1)$ transform the inversion of the chord into its complement.

Finally we need to observe that the inversion argument above will take care of the two remaining situations: $x_3 < 2 + p$ and $x_2 > 2$.

### B. Non-consecutive ones.

Notice that the case $p = 1$ was done above. Assume that the chord $P$ does not contain twos.

$$P = (x_1, 1, x_2, 1, x_3) \text{ with } P^C = (\overbrace{1,1,\ldots,1}^{x_1-2}, 3, \overbrace{1,1,\ldots,1}^{x_2-2}, 3, \overbrace{1,1,\ldots,1}^{x_3-2}, 2) \text{ and then:}$$

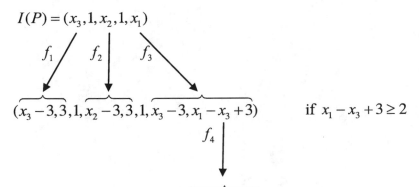

$$(x_3-3,3,1,\overbrace{x_2-3,3,1},\overbrace{x_3-3,x_1-x_3+3}) \qquad \text{if } x_1-x_3+3 \ge 2$$

$$f_4 \downarrow$$

$$(x_3-3,3,1,x_2-3,3,1,x_3-3,\overbrace{2,x_1-x_3+1}) \quad \text{and converting } x_3-3, x_2-3, x_3-3 \text{ and } x_1-x_3+1$$

to ones, we arrive at a representative of $\bar{P}^C$ after $12-2n$ enrichments.

If $x_1-x_3+3 < 2$ the chord must contain at least one two.

If in fact $P$ contains exactly one two, for instance $x_1 = 2$, then our chord $P = (2,1,x_2,1,x_3)$ will

have $P^C = (3,\overbrace{1,1,...,1}^{x_2-2},3,\overbrace{1,1,...,1}^{x_3-2},2)$ as a complement. Consider firstly $\qquad x_3 \ge 5$, then

$$I(P) = (x_3,1,x_2,1,2)$$

$$f_1 \swarrow \qquad f_2 \downarrow$$

$$(\overbrace{x_3-3,3,1},\overbrace{x_2-3,3,1},2)$$

$$f_3 \downarrow \qquad\qquad f_4 \downarrow$$

$$(\overbrace{x_3-5,2,3,1},x_2-3,3,1,\overbrace{1,1}) \text{ and converting } x_3-5 \text{ and } x_2-3 \text{ to ones we obtain a representative}$$

of $\bar{P}^C$ after $4+x_3-6+x_2-4 = 12-2n$ enrichments.

Once again if $x_3 < 4$, the inversion argument used in part A will convert this situation to the one above.

If $x_3 = 4$ then $P = (2,1,4,1,4)$ with $P^C = (3,1,1,3,1,1,2)$

and $f_4^3 \circ f_1^3(4,1,4,1,2) = (3,1,1,3,1,1,2) = P^C$.

Finally if we have two twos in the chord, then either $P = (2,1,2,1,6)$ with $P^C = (3,3,1,1,1,1,2)$ and $f_4^1 \circ f_1^3(6,1,2,1,2) = (3,3,1,1,1,1,2) = P^C$ or $P = (2,1,6,1,2)$ which belongs to $\overline{P*}$ and clearly we can see that it is impossible to find enrichments leading to $P^C = (3,1,1,1,1,3,2)$.

*CASE 4.* Chords with four intervals different from one.

The number of ones in the chord is $p \leq 1$. Thus we have two choices:

A. $p = 0$.

Assume first that $P$ does not contain twos.

Then $P = (3,3,3,3)$ with $P^C = (1,2,1,2,1,2,1,2)$.

And $I(P) = (3,3,3,3)$

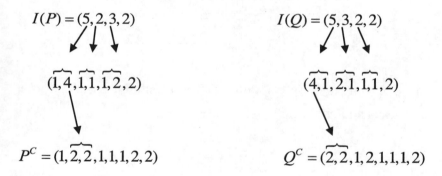

$(\overbrace{1,2},\overbrace{1,2},\overbrace{1,2},\overbrace{1,2})$ after four enrichments.

If $P$ contains one two, we have ( except for inversions) only two choices:

$P = (3,2,3,4)$ and $Q = (3,3,2,4)$. Below we display convenient enrichments.

$I(P) = (4,3,2,3)$

$P^C = (\overbrace{2,2},\overbrace{1,2},\overbrace{1,1},\overbrace{2,1})$

$I(Q) = (4,2,3,3)$

$Q^C = (\overbrace{2,2},\overbrace{1,1},\overbrace{2,1},\overbrace{2,1})$

Now if $P$ contains two twos, we have again ( except for inversions) two choices:

$P = (2,3,2,5)$ and $Q = (2,2,3,5)$ with enrichments:

$I(P) = (5,2,3,2)$

$(\overbrace{1,4},\overbrace{1,1},\overbrace{1,2},2)$

$P^C = (1,\overbrace{2,2},1,1,1,2,2)$

$I(Q) = (5,3,2,2)$

$(\overbrace{4,1},\overbrace{2,1},\overbrace{1,1},2)$

$Q^C = (\overbrace{2,2},1,2,1,1,1,2)$

If $P$ contains three twos, we have one choice: $P = (2,2,2,6)$ and
$f_3^1 \circ f_1^1 \circ f_1^2 \circ f_1^4 (6,2,2,2) = (1,1,1,1,2,2,2,2) = P^C$

B. $p = 1$.

Since the chord $P$ must contain at least one two, we have two possibilities:

$x_1 = 2$ (the case $x_4 = 2$ is an identical one by the inversion argument) and

$x_2 = 2$ (with $x_3 = 2$ as the inverted case).

Since the proof of these cases mirror each other completely, we will cover only the first one.

Thus now $P = (1, 2, x_2, x_3, x_4)$ with $P^C = (2, \overbrace{1, 1, ..., 1}^{x_2-2}, 2, \overbrace{1, 1, ..., 1}^{x_3-2}, 2, \overbrace{1, 1, ..., 1}^{x_4-2}, 3)$.

At least one of the $x_i$ must be greater than or equal to three.

Assume first that $\boxed{x_4 \geq 3}$

$I(P) = (x_4, x_3, x_2, 2, 1)$ with enrichments:

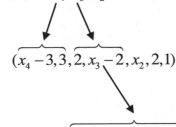

$(\overbrace{x_4 - 3, 3}, \overbrace{2, x_3 - 2}, x_2, 2, 1)$

$(x_4 - 3, 3, 2, \overbrace{x_2 - 2, x_3 - 2 - x_2 + 2}, x_2, 2, 1)$ and if $x_3 \geq x_2 + 2$

$(x_4 - 3, 3, 2, x_2 - 2, \overbrace{2, x_3 - x_2 - 2}, x_2, 2, 1)$ and converting $x_4 - 3, x_2 - 2, x_3 - x_2 - 2$ and $x_2$ to ones, we arrive at a representative of $\overline{P^C}$ after $12 - 2n$ enrichments.

If $x_3 = x_2$ above then, we proceed as follows after the first two enrichments:

$(x_4 - 3, 3, 2, x_3 - 2, x_2, 2, 1)$

$(x_4 - 3, 3, 2, x_3 - 2, \overbrace{2, x_2 - 2}, 2, 1)$ and once again we obtain a representative of $\overline{P^C}$ after converting $x_4 - 3, x_3 - 2$ and $x_2 - 2$ to ones and applying $12 - 2n$ enrichments.

If $x_3 = x_2 + 1$, we have only one choice:

$P = (1, 2, 2, 3, 4)$ with $P^C = (2, 2, 1, 2, 1, 1, 3)$ and $f_3^1 \circ f_1^2 (4, 3, 2, 2, 1) = (2, 1, 1, 3, 2, 2, 1) \in \overline{P^C}$.

If $x_3 < x_2$ we have two choices:

$P = (1, 2, 3, 2, 4)$ with $P^C = (2, 1, 2, 2, 1, 1, 3)$ and $Q = (1, 2, 4, 2, 3)$ with $Q^C = (2, 1, 1, 2, 2, 1, 3)$.

Now $f_1^2 \circ f_2^1 (4, 2, 3, 2, 1) = (2, 2, 1, 1, 3, 2, 1) \in \overline{P^C}$, and also

$f_3^1 \circ f_3^2 (3, 2, 4, 2, 1) = (3, 2, 1, 1, 2, 2, 1) \in \overline{Q^C}$.

Finally if $\boxed{x_4 = 2}$, we have, except for inversions, two choices left:

$P = (1, 2, 3, 4, 2)$ with $P^C = (2, 1, 2, 1, 1, 2, 3)$ and $Q = (1, 2, 5, 2, 2)$ with $Q^C = (2, 1, 1, 1, 2, 2, 3)$.

Notice that $f_2^1 \circ f_2^2(2,4,3,2,1) = (2,1,1,2,3,2,1) \in \overline{P^C}$ ,and also

$f_3^3 \circ f_4^1(2,2,5,2,1) = (2,2,3,2,1,1,1) \in \overline{Q^C}$ .

*CASE 5.* Five intervals different from one.

We have only three possibilities:

$P = (2,2,2,2,4)$ with $P^C = (2,2,2,2,1,1,2)$

$Q = (2,2,2,3,3)$ with $Q^C = (2,2,2,1,2,1,2)$ and

$R = (2,2,3,2,3)$ with $R^C = (2,2,1,2,2,1,2)$ .

Now $f_1^2 \circ f_4^1(4,2,2,2,2) = (2,2,2,2,1,1,2) = P^C$

$f_1^1 \circ f_2^1(3,3,2,2,2) = (1,2,1,2,2,2,2) \in \overline{Q^C}$ and

$f_1^1 \circ f_3^1(3,2,3,2,2) = (1,2,2,1,2,2,2) \in \overline{R^C}$ .

And the proof of (5.2) is completed.

We now introduce a dual result:

Given a chord $P$ in $\mathcal{T}^n$ , where $n > 6$, and $P$ is not in the orbit $\overline{P*^C}$ ,we can find a representative $P^C$ of $\overline{P^C}$ and $2n - 12$ reductions such that

$$P^C = \pi_k \circ \pi_{k-1} \circ \cdots \circ \pi_1 \circ I(P) \text{ , or } I(P^C) = \pi_k \circ \pi_{k-1} \circ \cdots \circ \pi_1(P) \quad \text{where } k = 2n - 12 \quad (5.3)$$

Property (5.3) derives from (5.2) in the following manner:

If $P$ is in $\mathcal{T}^n$ , where $n > 6$, $P^C$ is in $\mathcal{T}^{12-n}$ and since $12 - n < 6$, we can apply (5.2) to $P^C$ by finding $12 - 2(12 - n) = 2n - 12$ enrichments, such that

$$P = \left(P^C\right)^C = f_k \circ f_{k-1} \circ \cdots \circ f_1 \circ I(P^C)$$

For each $1 \le i \le k$ we can find a reduction $\pi_i$ such that $\pi_i \circ f_i =$ Identity , and thus:

$$\pi_k \circ \pi_{k-1} \circ \cdots \circ \pi_1(P) = I(P^C)$$

which is the second form of (5.3).

**Example 5.3**

$$P = (1, 1, 1, 1, 2, 1, 5) \in \mathcal{T}^7, \text{ with } P^C = (3, 1, 1, 1, 6) \in \mathcal{T}^5$$

In order to construct the complement, first take the inversion of $P$,

$$I(P) = I(1, 1, 1, 1, 2, 1, 5) = (5, 1, 2, 1, 1, 1, 1)$$

then, perform the following compositions of reductions:

$$P^C = \quad (\underbrace{5, 1}, 2, 1, 1, 1, 1)$$

$$\pi_{\#1} \downarrow$$

$$(6, \underbrace{2, 1}, 1, 1, 1)$$

$$\pi_{\#2} \downarrow$$

$$(6, 3, 1, 1, 1) = \sigma^4 (P^C)$$

Again, as was the case in the previous example, the initial inversion is necessary in this case as well.

---

Given a chord $P$ in $\mathcal{T}^6$ and any reduction $\pi_i$, $1 \leq i \leq 6$, not leading to a representative of $\overline{P*}$, we can find at most two enrichments $f_\ell$ and $f_k$ and one more reduction $\pi_j$ such that

$$I(P^C) = \pi_j \circ f_\ell \circ f_k \circ \pi_i (P) \qquad \text{where } I(P^C) \text{ notates a representative of } I(\overline{P}^C). \quad (5.4)$$

---

Property (5.4) also derives from (5.3):

Let's first prove that given any chord $P$ in $\mathcal{T}^n$ and any reduction $\pi_i$ we can find an enrichment $f_j$ such that

$$\mathcal{C}(\pi_i(P)) = f_j(\mathcal{C}(P)) \qquad (5.5)$$

Let $P = (x_1, \cdots, x_i, x_{i+1}, \cdots, x_n)$ and $\pi_i(P) = (x_1, \cdots, x_i + x_{i+1}, \cdots, x_n)$.

Let $p$ = The number of consecutive ones immediately before $x_i$, and

$p'$ = number of consecutive ones immediately after $x_{i+1}$.

Consider three different situations.

1) $x_i \neq 1$ and $x_{i+1} \neq 1$

$$\mathcal{C}(P) = \left( \overbrace{1, 1, \cdots, 1}^{x_1-2}, 2+p, \overbrace{1, 1, \cdots, 1}^{x_i-2}, 2, \overbrace{1, 1, \cdots, 1}^{x_{i+1}-2}, 2+p', \overbrace{1, 1, \cdots, 1}^{x_n-2}, 2 \right)$$

$$f_i \downarrow$$

$$\mathcal{C}(\pi_i(P)) = \left( \overbrace{1, 1, \cdots, 1}^{x_1-2}, 2+p, \underbrace{1, 1, \cdots, 1, \overbrace{1, 1}, 1, 1, \cdots, 1}_{x_i+x_{i+1}-2}, 2+p',, \overbrace{1, 1, \cdots, 1}^{x_n-2}, 2 \right)$$

2) $x_i = 1$ and $x_{i+1} = 1$

$$\mathcal{C}(P) = \left( \overbrace{1, 1, \cdots, 1}^{x_1-2}, 2+p+2+p', \overbrace{1, 1, \cdots, 1}^{x_n-2}, 2 \right)$$

$$f_i \downarrow$$

$$\mathcal{C}(\pi_i(P)) = \left( \overbrace{1, 1, \cdots, 1}^{x_1-2}, 2+p, \; 2+p', \overbrace{1, 1, \cdots, 1}^{x_n-2}, 2 \right)$$

3) $x_i = 1$ and $x_{i+1} \neq 1$   ($x_i \neq 1$ and $x_{i+1} = 1$ is identical)

$$\mathcal{C}(P) = \left( \overbrace{1, 1, \cdots, 1}^{x_1-2}, 2+p+1, \overbrace{1, 1, \cdots, 1}^{x_{i+1}-2}, 2+p', \overbrace{1, 1, \cdots, 1}^{x_n-2}, 2 \right)$$

$$f_i \downarrow$$

$$\mathcal{C}(\pi_i(P)) = \left( \overbrace{1, 1, \cdots, 1}^{x_1-2}, 2+p, 1, \overbrace{1, 1, \cdots, 1}^{x_{i+1}-2}, 2+p', \overbrace{1, 1, \cdots, 1}^{x_n-2}, 2 \right)$$

and (5.5) is proven.

Let's continue with the proof of (5.4).
Consider $P$ a hexachord in $\mathcal{T}^6$ and any reduction $\pi_i$ acting on $P$. From (5.5) we can find a $f_{j'}$ such that:

$$I\mathcal{C}(\pi_i(P)) = I f_{j'}(\mathcal{C}(P))$$

Recall from (3.3), we are guaranteed the existence of another enrichment $f_j$ that commutes with $I$:

$$I \circ f_{j'} = f_j \circ I$$

Thus, we can find a new enrichment $f_j$ such that

$$I\mathcal{C}\big(\pi_i(P)\big) = f_j \circ I\big(\mathcal{C}(P)\big)$$

Since $\pi_i(P) \in \mathcal{T}^5$, and $\pi_i(P) \notin \overline{P*}$, applying (5.2) we can find two enrichments $f_\ell$ and $f_k$ such that

$$I\mathcal{C}\big(\pi_i(P)\big) = f_\ell \circ f_k \circ \pi_i(P) = f_j \circ I\big(\mathcal{C}(P)\big)$$

Applying $\pi_j$ (inverse of $f_j$) on the left hand side of the last two terms above we get our result:

$$\pi_j \circ f_\ell \circ f_k \circ \pi_i(P) = I\big(P^C\big)$$

The "at most" in the statement of (5.4) is clarified by the following examples:

## Example 5.4

If $P = (3, 1, 2, 1, 1, 4)$, with $P^C = (1, 3, 4, 1, 1, 2)$

Notice that in this case $\sigma^4 \circ I\big(P^C\big) = P$ and if our goal is to get from $P$ to $P^C$ we do not need any enrichments or reductions to accomplish this. However if we take any reduction $\pi_i$ on $P$ all we need is one enrichment (namely the inverse of $\pi_i$) to arrive at a representative of $I\big(P^C\big)$.

## Example 5.5

If $P = (1, 2, 1, 1, 1, 6)$, with $P^C = (5, 1, 1, 1, 1, 3)$

If we choose $\pi_1(P) = (3, 1, 1, 1, 6)$ we only need one enrichment $f_5^1$ to arrive at a representative of $I\big(P^C\big)$ effectively:

$$f_1^5 \circ \pi_1(P) = (3, 1, 1, 1, 1, 5) = I\big(P^C\big)$$

If we choose $\pi_2(P) = (1, 3, 1, 1, 6)$ we need two enrichments $f_2^2$ and $f_6^1$ and a reduction $\pi_1$ to reach a representative of $I\big(P^C\big)$. In fact:

$$\pi_1 \circ f_6^1 \circ f_2^2 \circ \pi_1(P) \;=\; \pi_1 \circ f_6^1 \circ f_2^2(1, 3, 1, 1, 6) \;=\; \pi_1 \circ f_6^1(1, 2, 1, 1, 1, 6)$$

$$=\; \pi_1(1, 2, 1, 1, 1, 1, 5) \;=\; (3, 1, 1, 1, 1, 5) \;=\; I\left(P^C\right)$$

Now we arrive at the fourth result:

---

Given a chord $P$ in $\mathcal{T}^6$ and any well defined enrichment $f_i$ on $P$, not leading to a representative

of $\overline{P*^C}$ ,we can find at most two reductions $\pi_\ell$, $\pi_k$ and one more enrichment $f_j$ such that

$$I\left(P^C\right) = f_j \circ \pi_\ell \circ \pi_k \circ f_i(P) \qquad\qquad \text{where } I\left(P^C\right) \text{ notates a representative of } I\left(\overline{P}^{\,C}\right). \qquad (5.6)$$

---

(5.6) derives from (5.3):

Consider a hexachord $P$ in $\mathcal{T}^6$ and any one of the six possible enrichments $f_i$ defined on $P$.

Since $\pi_i \circ f_i(P) = P$ and thus: $\mathcal{C}\left(\pi_i\left(f_i(P)\right)\right) = \mathcal{C}(P)$. Applying (5.4) we can find an enrichment

$f_h$ such that $f_h\left(\mathcal{C}\left(f_i(P)\right)\right) = \mathcal{C}(P)$ and also $I \circ f_h\left(\mathcal{C}\left(f_i(P)\right)\right) = I(P^C)$.

By (3.3) we can select an enrichment $f_j$ such that $f_j \circ I\left(\mathsf{C}\left(f_i(P)\right)\right) = I(P^C)$.

Since $f_i(P) \in \mathcal{T}^7$, and $f_i(P) \notin \overline{P*^C}$, applying (5.3) we can find reductions $\pi_\ell$ and $\pi_k$ such that:

$$f_j \circ \pi_\ell \circ \pi_k \circ f_i(P) = I\left(P^C\right).$$

## Example 5.6

Let $P = (2, 1, 2, 1, 1, 5)$, with $P^C = (3, 4, 1, 1, 1, 2)$.

If we choose $f_6^1$ then all we need is reduction $\pi_1$ to get to a representation of $I\left(\overline{P^C}\right)$:

$$\pi_1 \circ f_6^1(2, 1, 2, 1, 1, 5) = \pi_1(2, 1, 2, 1, 1, 1, 4) = (3, 2, 1, 1, 1, 4) = \sigma^5 \circ I(P^C)$$

If we select $f_3^1$ then we need two reductions $\pi_5$, $\pi_5$ and enrichment $f_4^5$ to get to a representative

of $I\left(\overline{P^C}\right)$:

$$f_5^4 \circ \pi_5 \circ \pi_5 \circ f_3^1(2, 1, 2, 1, 1, 5) = f_5^4 \circ \pi_5 \circ \pi_5(2, 1, 1, 1, 1, 1, 5)$$

$$= f_5^4 \circ \pi_5(2, 1, 1, 1, 2, 5) = f_5^4(2, 1, 1, 1, 7) = (2, 1, 1, 1, 4, 3) = I(P^C)$$

# Appendix A. Table of $n$-chords

The $n$-chords in interval notation $P = (x_1, x_2, x_3, \cdots, x_n)$ are sorted by Index $\|P\|$, then by the Hankel matrix $\left|\det(M_H)\right|$, and $\mathrm{rank}(M_H)$. The corresponding mesh $\widehat{\sigma_k}$, and the $I$-symmetry index are listed.

(The "prime form" notation is constructed by ordering the intervals in decreasing order from right to left and it is consistent with the Rahn/Morris refinement of the Forte notation.( [1], [5] and [7] ). Inversions of the chord are denoted by the Forte label of the chord followed by the symbol $I$ )

## $n = 2$

| Prime form | Forte Name | Index $\|P\|$ | $\left|\det(M_H)\right|$ | $\mathrm{rank}(M_H)$ | $\widehat{\sigma_1}$ | $I$–sym |
|---|---|---|---|---|---|---|
| (1, 11) | 2–01 | 5 | 120 | 2 | 10 | 2 |
| (2, 10) | 2–02 | 4 | 96 | 2 | 8 | 2 |
| (3, 9) | 2–03 | 3 | 72 | 2 | 6 | 2 |
| (4, 8) | 2–04 | 2 | 48 | 2 | 4 | 2 |
| (5, 7) | 2–05 | 1 | 24 | 2 | 2 | 2 |
| (6, 6) | 2–06 | 0 | 0 | 1 | 0 | 0 |

## $n = 3$

| Prime form | Forte Name | Index $\|P\|$ | $\left|\det(M_H)\right|$ | $\mathrm{rank}(M_H)$ | $\widehat{\sigma_1}$ | $I$–sym |
|---|---|---|---|---|---|---|
| (1, 1, 10) | 3–01 | $3\sqrt{3}$ | 972 | 3 | 3 | 0 |
| (1, 2, 9) | 3–02 | $\sqrt{19}$ | 684 | 3 | $\sqrt{57}$ | $1/\sqrt{19}$ |
| (2, 1, 9) | 3–02 $I$ | $\sqrt{19}$ | 684 | 3 | $\sqrt{57}$ | $1/\sqrt{19}$ |
| (1, 3, 8) | 3–03 | $\sqrt{13}$ | 468 | 3 | $\sqrt{39}$ | $2/\sqrt{13}$ |
| (3, 1, 8) | 3–03 $I$ | $\sqrt{13}$ | 468 | 3 | $\sqrt{39}$ | $2/\sqrt{13}$ |
| (2, 2, 8) | 3–06 | $2\sqrt{3}$ | 432 | 3 | 6 | 0 |
| (1, 4, 7) | 3–04 | 3 | 324 | 3 | $3\sqrt{3}$ | 1 |
| (4, 1, 7) | 3–04 $I$ | 3 | 324 | 3 | $3\sqrt{3}$ | 1 |
| (1, 5, 6) | 3–05 | $\sqrt{7}$ | 252 | 3 | $\sqrt{21}$ | $1/\sqrt{7}$ |
| (5, 1, 6) | 3–05 $I$ | $\sqrt{7}$ | 252 | 3 | $\sqrt{21}$ | $1/\sqrt{7}$ |
| (2, 3, 7) | 3–07 | $\sqrt{7}$ | 252 | 3 | $\sqrt{21}$ | $1/\sqrt{7}$ |
| (3, 2, 7) | 3–07 $I$ | $\sqrt{7}$ | 252 | 3 | $\sqrt{21}$ | $1/\sqrt{7}$ |
| (2, 4, 6) | 3–08 | 2 | 144 | 3 | $2\sqrt{3}$ | 1 |
| (4, 2, 6) | 3–08 $I$ | 2 | 144 | 3 | $2\sqrt{3}$ | 1 |
| (2, 5, 5) | 3–09 | $\sqrt{3}$ | 108 | 3 | 3 | 0 |
| (3, 3, 6) | 3–10 | $\sqrt{3}$ | 108 | 3 | 3 | 0 |
| (3, 4, 5) | 3–11 | 1 | 36 | 3 | $\sqrt{3}$ | 1 |
| (4, 3, 5) | 3–11 $I$ | 1 | 36 | 3 | $\sqrt{3}$ | 1 |
| (4, 4, 4) | 3–12 | 0 | 0 | 1 | 0 | 0 |

**n = 4**

| Prime form | Forte Name | Index $\|P\|$ | $\|\det(M_H)\|$ | rank$(M_H)$ | $\widehat{\sigma_1}$ | $\widehat{\sigma_2}$ | $I$–sym |
|---|---|---|---|---|---|---|---|
| (1, 1, 1, 9) | 4-01 | $2\sqrt{6}$ | 6144 | 4 | 8 | 8 | 0 |
| (1, 1, 2, 8) | 4-02 | $\sqrt{17}$ | 3600 | 4 | $\sqrt{43}$ | $5\sqrt{2}$ | $1/\sqrt{17}$ |
| (2, 1, 1, 8) | 4-02 $I$ | $\sqrt{17}$ | 3600 | 4 | $\sqrt{43}$ | $5\sqrt{2}$ | $1/\sqrt{17}$ |
| (1, 2, 1, 8) | 4-03 | $\sqrt{17}$ | 3456 | 4 | $5\sqrt{2}$ | 6 | 0 |
| (1, 1, 3, 7) | 4-04 | $2\sqrt{3}$ | 1920 | 4 | $2\sqrt{7}$ | $2\sqrt{10}$ | $1/\sqrt{3}$ |
| (3, 1, 1, 7) | 4-04 $I$ | $2\sqrt{3}$ | 1920 | 4 | $2\sqrt{7}$ | $2\sqrt{10}$ | $1/\sqrt{3}$ |
| (1, 3, 1, 7) | 4-07 | $2\sqrt{3}$ | 1536 | 4 | $2\sqrt{10}$ | 4 | 0 |
| (1, 2, 2, 7) | 4-11 | $\sqrt{11}$ | 1872 | 4 | $\sqrt{31}$ | $\sqrt{26}$ | $1/\sqrt{11}$ |
| (2, 2, 1, 7) | 4-11 $I$ | $\sqrt{11}$ | 1872 | 4 | $\sqrt{31}$ | $\sqrt{26}$ | $1/\sqrt{11}$ |
| (2, 1, 2, 7) | 4-10 | $\sqrt{11}$ | 1728 | 4 | $\sqrt{26}$ | 6 | 0 |
| (1, 1, 4, 6) | 4-05 | 3 | 816 | 4 | $\sqrt{19}$ | $\sqrt{34}$ | 2/3 |
| (4, 1, 1, 6) | 4-05 $I$ | 3 | 816 | 4 | $\sqrt{19}$ | $\sqrt{34}$ | 2/3 |
| (1, 4, 1, 6) | 4-08 | 3 | 384 | 4 | $\sqrt{34}$ | 2 | 0 |
| (1, 1, 5, 5) | 4-06 | $2\sqrt{2}$ | 0 | 3 | 4 | $4\sqrt{2}$ | 0 |
| (1, 5, 1, 5) | 4-09 | $2\sqrt{2}$ | 0 | 2 | $4\sqrt{2}$ | 0 | 0 |
| (1, 2, 3, 6) | 4-13 | $\sqrt{7}$ | 960 | 4 | $3\sqrt{2}$ | $2\sqrt{5}$ | $2/\sqrt{7}$ |
| (3, 2, 1, 6) | 4-13 $I$ | $\sqrt{7}$ | 960 | 4 | $3\sqrt{2}$ | $2\sqrt{5}$ | $2/\sqrt{7}$ |
| (1, 3, 2, 6) | 4-Z15 | $\sqrt{7}$ | 720 | 4 | $\sqrt{23}$ | $\sqrt{10}$ | $1/\sqrt{7}$ |
| (2, 3, 1, 6) | 4-Z15 $I$ | $\sqrt{7}$ | 720 | 4 | $\sqrt{23}$ | $\sqrt{10}$ | $1/\sqrt{7}$ |
| (2, 1, 3, 6) | 4-12 | $\sqrt{7}$ | 624 | 4 | $\sqrt{15}$ | $\sqrt{26}$ | $1/\sqrt{7}$ |
| (3, 1, 2, 6) | 4-12 $I$ | $\sqrt{7}$ | 624 | 4 | $\sqrt{15}$ | $\sqrt{26}$ | $1/\sqrt{7}$ |
| (2, 2, 2, 6) | 4-21 | $\sqrt{6}$ | 768 | 4 | 4 | 4 | 0 |
| (1, 2, 4, 5) | 4-Z29 | $\sqrt{5}$ | 432 | 4 | $\sqrt{11}$ | $3\sqrt{2}$ | $\sqrt{2/5}$ |
| (4, 2, 1, 5) | 4-Z29 $I$ | $\sqrt{5}$ | 432 | 4 | $\sqrt{11}$ | $3\sqrt{2}$ | $\sqrt{2/5}$ |
| (1, 4, 2, 5) | 4-16 | $\sqrt{5}$ | 144 | 4 | $\sqrt{19}$ | $\sqrt{2}$ | $1/\sqrt{5}$ |
| (2, 4, 1, 5) | 4-16 $I$ | $\sqrt{5}$ | 144 | 4 | $\sqrt{19}$ | $\sqrt{2}$ | $1/\sqrt{5}$ |
| (2, 1, 4, 5) | 4-14 | $\sqrt{5}$ | 0 | 3 | $\sqrt{10}$ | $2\sqrt{5}$ | $\sqrt{2/5}$ |
| (4, 1, 2, 5) | 4-14 $I$ | $\sqrt{5}$ | 0 | 3 | $\sqrt{10}$ | $2\sqrt{5}$ | $\sqrt{2/5}$ |
| (1, 3, 3, 5) | 4-18 | 2 | 384 | 4 | $2\sqrt{3}$ | $2\sqrt{2}$ | 1 |
| (3, 3, 1, 5) | 4-18 $I$ | 2 | 384 | 4 | $2\sqrt{3}$ | $2\sqrt{2}$ | 1 |
| (3, 1, 3, 5) | 4-17 | 2 | 0 | 3 | $2\sqrt{2}$ | 4 | 0 |
| (1, 3, 4, 4) | 4-19 | $\sqrt{3}$ | 240 | 4 | $\sqrt{7}$ | $\sqrt{10}$ | $1/\sqrt{3}$ |
| (3, 1, 4, 4) | 4-19 $I$ | $\sqrt{3}$ | 240 | 4 | $\sqrt{7}$ | $\sqrt{10}$ | $1/\sqrt{3}$ |
| (2, 2, 3, 5) | 4-22 | $\sqrt{3}$ | 240 | 4 | $\sqrt{7}$ | $\sqrt{10}$ | $1/\sqrt{3}$ |
| (3, 2, 2, 5) | 4-22 $I$ | $\sqrt{3}$ | 240 | 4 | $\sqrt{7}$ | $\sqrt{10}$ | $1/\sqrt{3}$ |
| (2, 3, 2, 5) | 4-23 | $\sqrt{3}$ | 192 | 4 | $\sqrt{10}$ | 2 | 0 |
| (1, 4, 3, 4) | 4-20 | $\sqrt{3}$ | 192 | 4 | $\sqrt{10}$ | 2 | 0 |
| (2, 2, 4, 4) | 4-24 | $\sqrt{2}$ | 0 | 3 | 2 | $2\sqrt{2}$ | 0 |
| (2, 4, 2, 4) | 4-25 | $\sqrt{2}$ | 0 | 2 | $2\sqrt{2}$ | 0 | 0 |
| (2, 3, 3, 4) | 4-27 | 1 | 48 | 4 | $\sqrt{3}$ | $\sqrt{2}$ | 1 |
| (3, 3, 2, 4) | 4-27 $I$ | 1 | 48 | 4 | $\sqrt{3}$ | $\sqrt{2}$ | 1 |
| (3, 2, 3, 4) | 4-26 | 1 | 0 | 3 | $\sqrt{2}$ | 2 | 0 |
| (3, 3, 3, 3) | 4-28 | 0 | 0 | 1 | 0 | 0 | 0 |

**n = 5**

| Prime form | Forte Name | $\sqrt{5}\lVert P\rVert$ | $\lvert\det(M_H)\rvert$ | $\text{rank}(M_H)$ | $\widehat{\sigma_1}$ | $\widehat{\sigma_2}$ | I–sym |
|---|---|---|---|---|---|---|---|
| (1,1,1,1,8) | 5-01 | $7\sqrt{2}$ | 28812 | 5 | 7 | 7 | 0 |
| (1,1,1,2,7) | 5-02 | $2\sqrt{17}$ | 13332 | 5 | $\sqrt{31}$ | $\sqrt{37}$ | $\sqrt{5}/(2\sqrt{17})$ |
| (2,1,1,1,7) | 5-02 I | $2\sqrt{17}$ | 13332 | 5 | $\sqrt{31}$ | $\sqrt{37}$ | $\sqrt{5}/(2\sqrt{17})$ |
| (1,1,2,1,7) | 5-03 | $2\sqrt{17}$ | 13332 | 5 | $\sqrt{37}$ | $\sqrt{31}$ | $\sqrt{5}/(2\sqrt{17})$ |
| (1,2,1,1,7) | 5-03 I | $2\sqrt{17}$ | 13332 | 5 | $\sqrt{37}$ | $\sqrt{31}$ | $\sqrt{5}/(2\sqrt{17})$ |
| (1,1,1,3,6) | 5-04 | $4\sqrt{3}$ | 5412 | 5 | $\sqrt{19}$ | $\sqrt{29}$ | $\sqrt{5}/(2\sqrt{3})$ |
| (3,1,1,1,6) | 5-04 I | $4\sqrt{3}$ | 5412 | 5 | $\sqrt{19}$ | $\sqrt{29}$ | $\sqrt{5}/(2\sqrt{3})$ |
| (1,1,3,1,6) | 5-06 | $4\sqrt{3}$ | 5412 | 5 | $\sqrt{29}$ | $\sqrt{19}$ | $\sqrt{5}/(2\sqrt{3})$ |
| (1,3,1,1,6) | 5-06 I | $4\sqrt{3}$ | 5412 | 5 | $\sqrt{29}$ | $\sqrt{19}$ | $\sqrt{5}/(2\sqrt{3})$ |
| (1,1,2,2,6) | 5-09 | $\sqrt{43}$ | 5532 | 5 | $\sqrt{21}$ | $\sqrt{22}$ | $\sqrt{10/43}$ |
| (2,2,1,1,6) | 5-09 I | $\sqrt{43}$ | 5532 | 5 | $\sqrt{21}$ | $\sqrt{22}$ | $\sqrt{10/43}$ |
| (1,2,1,2,6) | 5-10 | $\sqrt{43}$ | 5532 | 5 | $\sqrt{22}$ | $\sqrt{21}$ | $\sqrt{10/43}$ |
| (2,1,2,1,6) | 5-10 I | $\sqrt{43}$ | 5532 | 5 | $\sqrt{22}$ | $\sqrt{21}$ | $\sqrt{10/43}$ |
| (2,1,1,2,6) | 5-08 | $\sqrt{43}$ | 4332 | 5 | $\sqrt{17}$ | $\sqrt{26}$ | 0 |
| (1,2,2,1,6) | 5-Z12 | $\sqrt{43}$ | 4332 | 5 | $\sqrt{26}$ | $\sqrt{17}$ | 0 |
| (1,1,1,4,5) | 5-05 | $\sqrt{38}$ | 2172 | 5 | $\sqrt{13}$ | 5 | $\sqrt{5/38}$ |
| (4,1,1,1,5) | 5-05 I | $\sqrt{38}$ | 2172 | 5 | $\sqrt{13}$ | 5 | $\sqrt{5/38}$ |
| (1,1,4,1,5) | 5-07 | $\sqrt{38}$ | 2172 | 5 | 5 | $\sqrt{13}$ | $\sqrt{5/38}$ |
| (1,4,1,1,5) | 5-07 I | $\sqrt{38}$ | 2172 | 5 | 5 | $\sqrt{13}$ | $\sqrt{5/38}$ |
| (1,1,3,2,5) | 5-14 | $2\sqrt{7}$ | 2292 | 5 | $\sqrt{15}$ | $\sqrt{13}$ | $\sqrt{5/7}$ |
| (2,3,1,1,5) | 5-14 I | $2\sqrt{7}$ | 2292 | 5 | $\sqrt{15}$ | $\sqrt{13}$ | $\sqrt{5/7}$ |
| (1,2,1,3,5) | 5-16 | $2\sqrt{7}$ | 2292 | 5 | $\sqrt{13}$ | $\sqrt{15}$ | $\sqrt{5/7}$ |
| (3,1,2,1,5) | 5-16 I | $2\sqrt{7}$ | 2292 | 5 | $\sqrt{13}$ | $\sqrt{15}$ | $\sqrt{5/7}$ |
| (1,3,1,2,5) | 5-Z18 | $2\sqrt{7}$ | 1812 | 5 | $\sqrt{17}$ | $\sqrt{11}$ | $5/(2\sqrt{7})$ |
| (2,1,3,1,5) | 5-Z18 I | $2\sqrt{7}$ | 1812 | 5 | $\sqrt{17}$ | $\sqrt{11}$ | $5/(2\sqrt{7})$ |
| (1,1,2,3,5) | 5-Z36 | $2\sqrt{7}$ | 1812 | 5 | $\sqrt{11}$ | $\sqrt{17}$ | $5/(2\sqrt{7})$ |
| (3,2,1,1,5) | 5-Z36 I | $2\sqrt{7}$ | 1812 | 5 | $\sqrt{11}$ | $\sqrt{17}$ | $5/(2\sqrt{7})$ |
| (2,1,1,3,5) | 5-11 | $2\sqrt{7}$ | 852 | 5 | 3 | $\sqrt{19}$ | $\sqrt{5}/(2\sqrt{7})$ |
| (3,1,1,2,5) | 5-11 I | $2\sqrt{7}$ | 852 | 5 | 3 | $\sqrt{19}$ | $\sqrt{5}/(2\sqrt{7})$ |
| (1,2,3,1,5) | 5-19 | $2\sqrt{7}$ | 852 | 5 | $\sqrt{19}$ | 3 | $\sqrt{5}/(2\sqrt{7})$ |
| (1,3,2,1,5) | 5-19 I | $2\sqrt{7}$ | 852 | 5 | $\sqrt{19}$ | 3 | $\sqrt{5}/(2\sqrt{7})$ |
| (1,1,4,2,4) | 5-15 | $\sqrt{23}$ | 1452 | 5 | $\sqrt{13}$ | $\sqrt{10}$ | 0 |
| (1,2,1,4,4) | 5-Z17 | $\sqrt{23}$ | 1452 | 5 | $\sqrt{10}$ | $\sqrt{13}$ | 0 |
| (2,1,2,2,5) | 5-23 | $\sqrt{23}$ | 1452 | 5 | $\sqrt{10}$ | $\sqrt{13}$ | $\sqrt{5/23}$ |
| (2,2,1,2,5) | 5-23 I | $\sqrt{23}$ | 1452 | 5 | $\sqrt{10}$ | $\sqrt{13}$ | $\sqrt{5/23}$ |
| (1,2,2,2,5) | 5-24 | $\sqrt{23}$ | 1452 | 5 | $\sqrt{13}$ | $\sqrt{10}$ | $\sqrt{5/23}$ |
| (2,2,2,1,5) | 5-24 I | $\sqrt{23}$ | 1452 | 5 | $\sqrt{13}$ | $\sqrt{10}$ | $\sqrt{5/23}$ |
| (1,1,2,4,4) | 5-13 | $\sqrt{23}$ | 372 | 5 | $\sqrt{7}$ | 4 | $\sqrt{5/23}$ |
| (2,1,1,4,4) | 5-13 I | $\sqrt{23}$ | 372 | 5 | $\sqrt{7}$ | 4 | $\sqrt{5/23}$ |
| (1,4,1,2,4) | 5-20 | $\sqrt{23}$ | 372 | 5 | 4 | $\sqrt{7}$ | $\sqrt{5/23}$ |
| (2,1,4,1,4) | 5-20 I | $\sqrt{23}$ | 372 | 5 | 4 | $\sqrt{7}$ | $\sqrt{5/23}$ |
| (1,3,1,3,4) | 5-21 | $3\sqrt{2}$ | 732 | 5 | $\sqrt{11}$ | $\sqrt{7}$ | $\sqrt{5}/(3\sqrt{2})$ |
| (3,1,3,1,4) | 5-21 I | $3\sqrt{2}$ | 732 | 5 | $\sqrt{11}$ | $\sqrt{7}$ | $\sqrt{5}/(3\sqrt{2})$ |

| Prime form | Forte Name | Index $\|P\|$ | $\det(M_H)$ | rank$(M_H)$ | $\widehat{\sigma_1}$ | $\widehat{\sigma_2}$ | $I$–sym |
|---|---|---|---|---|---|---|---|
| (1,1,3,3,4) | 5-Z38 | $3\sqrt{2}$ | 732 | 5 | $\sqrt{7}$ | $\sqrt{11}$ | $\sqrt{5}/(3\sqrt{2})$ |
| (3,3,1,1,4) | 5-Z38 $I$ | $3\sqrt{2}$ | 732 | 5 | $\sqrt{7}$ | $\sqrt{11}$ | $\sqrt{5}/(3\sqrt{2})$ |
| (1,3,3,1,4) | 5-22 | $3\sqrt{2}$ | 12 | 5 | $\sqrt{13}$ | $\sqrt{5}$ | 0 |
| (3,1,1,3,4) | 5-Z37 | $3\sqrt{2}$ | 12 | 5 | $\sqrt{5}$ | $\sqrt{13}$ | 0 |
| (1,2,2,3,4) | 5-27 | $\sqrt{13}$ | 492 | 5 | $\sqrt{6}$ | $\sqrt{7}$ | $\sqrt{10/13}$ |
| (3,2,2,1,4) | 5-27 $I$ | $\sqrt{13}$ | 492 | 5 | $\sqrt{6}$ | $\sqrt{7}$ | $\sqrt{10/13}$ |
| (2,1,3,2,4) | 5-28 | $\sqrt{13}$ | 492 | 5 | $\sqrt{7}$ | $\sqrt{6}$ | $\sqrt{10/13}$ |
| (2,3,1,2,4) | 5-28 $I$ | $\sqrt{13}$ | 492 | 5 | $\sqrt{7}$ | $\sqrt{6}$ | $\sqrt{10/13}$ |
| (2,2,1,3,4) | 5-26 | $\sqrt{13}$ | 372 | 5 | $\sqrt{5}$ | $2\sqrt{2}$ | $\sqrt{10/13}$ |
| (3,1,2,2,4) | 5-26 $I$ | $\sqrt{13}$ | 372 | 5 | $\sqrt{5}$ | $2\sqrt{2}$ | $\sqrt{10/13}$ |
| (1,2,3,2,4) | 5-29 | $\sqrt{13}$ | 372 | 5 | $2\sqrt{2}$ | $\sqrt{5}$ | $\sqrt{10/13}$ |
| (2,3,2,1,4) | 5-29 $I$ | $\sqrt{13}$ | 372 | 5 | $2\sqrt{2}$ | $\sqrt{5}$ | $\sqrt{10/13}$ |
| (2,1,2,3,4) | 5-25 | $\sqrt{13}$ | 132 | 5 | 2 | 3 | $\sqrt{5/13}$ |
| (3,2,1,2,4) | 5-25 $I$ | $\sqrt{13}$ | 132 | 5 | 2 | 3 | $\sqrt{5/13}$ |
| (1,3,2,2,4) | 5-30 | $\sqrt{13}$ | 132 | 5 | 3 | 2 | $\sqrt{5/13}$ |
| (2,2,3,1,4) | 5-30 $I$ | $\sqrt{13}$ | 132 | 5 | 3 | 2 | $\sqrt{5/13}$ |
| (2,2,2,2,4) | 5-33 | $2\sqrt{2}$ | 192 | 5 | 2 | 2 | 0 |
| (1,2,3,3,3) | 5-31 | $2\sqrt{2}$ | 132 | 5 | $\sqrt{3}$ | $\sqrt{5}$ | $\sqrt{5}/(2\sqrt{2})$ |
| (2,1,3,3,3) | 5-31 $I$ | $2\sqrt{2}$ | 132 | 5 | $\sqrt{3}$ | $\sqrt{5}$ | $\sqrt{5}/(2\sqrt{2})$ |
| (1,3,2,3,3) | 5-32 | $2\sqrt{2}$ | 132 | 5 | $\sqrt{5}$ | $\sqrt{3}$ | $\sqrt{5}/(2\sqrt{2})$ |
| (2,3,1,3,3) | 5-32 $I$ | $2\sqrt{2}$ | 132 | 5 | $\sqrt{5}$ | $\sqrt{3}$ | $\sqrt{5}/(2\sqrt{2})$ |
| (2,2,2,3,3) | 5-34 | $\sqrt{3}$ | 12 | 5 | 1 | $\sqrt{2}$ | 0 |
| (2,2,3,2,3) | 5-35 | $\sqrt{3}$ | 12 | 5 | $\sqrt{2}$ | 1 | 0 |

## $n = 6$

| Prime form | Forte Name | Index $\|P\|$ | $\det(M_H)$ | rank$(M_H)$ | $\widehat{\sigma_1}$ | $\widehat{\sigma_2}$ | $\widehat{\sigma_3}$ | $I$–sym |
|---|---|---|---|---|---|---|---|---|
| (1,1,1,1,1,7) | 6-01 | $\sqrt{15}$ | 93312 | 6 | 6 | 6 | 6 | 0 |
| (1,1,1,2,1,6) | 6-Z03 | $\sqrt{10}$ | 31752 | 6 | $\sqrt{26}$ | $\sqrt{21}$ | $\sqrt{26}$ | $1/\sqrt{10}$ |
| (1,2,1,1,1,6) | 6-Z03 $I$ | $\sqrt{10}$ | 31752 | 6 | $\sqrt{26}$ | $\sqrt{21}$ | $\sqrt{26}$ | $1/\sqrt{10}$ |
| (1,1,1,1,2,6) | 6-02 | $\sqrt{10}$ | 31248 | 6 | $\sqrt{21}$ | $\sqrt{26}$ | $\sqrt{26}$ | $1/\sqrt{10}$ |
| (2,1,1,1,1,6) | 6-02 $I$ | $\sqrt{10}$ | 31248 | 6 | $\sqrt{21}$ | $\sqrt{26}$ | $\sqrt{26}$ | $1/\sqrt{10}$ |
| (1,1,2,1,1,6) | 6-Z04 | $\sqrt{10}$ | 27648 | 6 | $\sqrt{26}$ | $\sqrt{26}$ | 4 | 0 |
| (1,1,1,3,1,5) | 6-05 | $\sqrt{7}$ | 10368 | 6 | $2\sqrt{5}$ | $2\sqrt{3}$ | $2\sqrt{5}$ | $2/\sqrt{7}$ |
| (1,3,1,1,1,5) | 6-05 $I$ | $\sqrt{7}$ | 10368 | 6 | $2\sqrt{5}$ | $2\sqrt{3}$ | $2\sqrt{5}$ | $2/\sqrt{7}$ |
| (1,1,1,1,3,5) | 6-Z36 | $\sqrt{7}$ | 8064 | 6 | $2\sqrt{3}$ | $2\sqrt{5}$ | $2\sqrt{5}$ | $2/\sqrt{7}$ |
| (3,1,1,1,1,5) | 6-Z36 $I$ | $\sqrt{7}$ | 8064 | 6 | $2\sqrt{3}$ | $2\sqrt{5}$ | $2\sqrt{5}$ | $2/\sqrt{7}$ |
| (1,1,3,1,1,5) | 6-Z06 | $\sqrt{7}$ | 3456 | 6 | $2\sqrt{5}$ | $2\sqrt{5}$ | 2 | 0 |
| (1,2,1,1,2,5) | 6-Z10 | $\sqrt{6}$ | 9216 | 6 | $\sqrt{14}$ | $\sqrt{14}$ | 4 | $1/\sqrt{3}$ |
| (2,1,1,2,1,5) | 6-Z10 $I$ | $\sqrt{6}$ | 9216 | 6 | $\sqrt{14}$ | $\sqrt{14}$ | 4 | $1/\sqrt{3}$ |
| (1,1,1,2,2,5) | 6-09 | $\sqrt{6}$ | 8208 | 6 | $\sqrt{13}$ | $\sqrt{14}$ | $3\sqrt{2}$ | $1/\sqrt{3}$ |
| (2,2,1,1,1,5) | 6-09 $I$ | $\sqrt{6}$ | 8208 | 6 | $\sqrt{13}$ | $\sqrt{14}$ | $3\sqrt{2}$ | $1/\sqrt{3}$ |
| (1,1,2,2,1,5) | 6-Z12 | $\sqrt{6}$ | 7056 | 6 | $\sqrt{17}$ | $\sqrt{14}$ | $\sqrt{10}$ | $1/\sqrt{6}$ |
| (1,2,2,1,1,5) | 6-Z12 $I$ | $\sqrt{6}$ | 7056 | 6 | $\sqrt{17}$ | $\sqrt{14}$ | $\sqrt{10}$ | $1/\sqrt{6}$ |
| (1,1,2,1,2,5) | 6-Z11 | $\sqrt{6}$ | 6552 | 6 | $\sqrt{14}$ | $\sqrt{17}$ | $\sqrt{10}$ | $1/\sqrt{6}$ |
| (2,1,2,1,1,5) | 6-Z11 $I$ | $\sqrt{6}$ | 6552 | 6 | $\sqrt{14}$ | $\sqrt{17}$ | $\sqrt{10}$ | $1/\sqrt{6}$ |

| | | | | | | | | |
|---|---|---|---|---|---|---|---|---|
| (1,2,1,2,1,5) | 6-Z13 | $\sqrt{6}$ | 5832 | 6 | $3\sqrt{2}$ | 3 | $3\sqrt{2}$ | 0 |
| (1,1,1,4,1,4) | 6-Z38 | $\sqrt{6}$ | 5832 | 6 | $3\sqrt{2}$ | 3 | $3\sqrt{2}$ | 0 |
| (2,1,1,1,2,5) | 6-08 | $\sqrt{6}$ | 5400 | 6 | $\sqrt{10}$ | $\sqrt{17}$ | $3\sqrt{2}$ | 0 |
| (1,1,1,1,4,4) | 6-Z37 | $\sqrt{6}$ | 0 | 5 | 3 | $3\sqrt{2}$ | $3\sqrt{2}$ | 0 |
| (1,1,4,1,1,4) | 6-07 | $\sqrt{6}$ | 0 | 3 | $3\sqrt{2}$ | $3\sqrt{2}$ | 0 | 0 |
| (1,3,1,1,2,4) | 6-16 | 2 | 3024 | 6 | $\sqrt{11}$ | $2\sqrt{2}$ | $\sqrt{10}$ | $1/\sqrt{2}$ |
| (2,1,1,3,1,4) | 6-16 $I$ | 2 | 3024 | 6 | $\sqrt{11}$ | $2\sqrt{2}$ | $\sqrt{10}$ | $1/\sqrt{2}$ |
| (1,2,1,1,3,4) | 6-14 | 2 | 2808 | 6 | $2\sqrt{2}$ | $\sqrt{11}$ | $\sqrt{10}$ | $1/\sqrt{2}$ |
| (3,1,1,2,1,4) | 6-14 $I$ | 2 | 2808 | 6 | $2\sqrt{2}$ | $\sqrt{11}$ | $\sqrt{10}$ | $1/\sqrt{2}$ |
| (1,1,2,3,1,4) | 6-Z17 | 2 | 2304 | 6 | $2\sqrt{3}$ | $2\sqrt{2}$ | $2\sqrt{2}$ | $1/\sqrt{2}$ |
| (1,3,2,1,1,4) | 6-Z17 $I$ | 2 | 2304 | 6 | $2\sqrt{3}$ | $2\sqrt{2}$ | $2\sqrt{2}$ | $1/\sqrt{2}$ |
| (1,1,1,3,2,4) | 6-Z41 | 2 | 1872 | 6 | 3 | $2\sqrt{2}$ | $\sqrt{14}$ | 1/2 |
| (2,3,1,1,1,4) | 6-Z41 $I$ | 2 | 1872 | 6 | 3 | $2\sqrt{2}$ | $\sqrt{14}$ | 1/2 |
| (1,1,1,2,3,4) | 6-Z40 | 2 | 1368 | 6 | $\sqrt{6}$ | $\sqrt{11}$ | $\sqrt{14}$ | $1/\sqrt{2}$ |
| (3,2,1,1,1,4) | 6-Z40 $I$ | 2 | 1368 | 6 | $\sqrt{6}$ | $\sqrt{11}$ | $\sqrt{14}$ | $1/\sqrt{2}$ |
| (1,2,1,3,1,4) | 6-Z19 | 2 | 648 | 6 | $\sqrt{14}$ | $\sqrt{3}$ | $\sqrt{14}$ | 1/2 |
| (1,3,1,2,1,4) | 6-Z19 $I$ | 2 | 648 | 6 | $\sqrt{14}$ | $\sqrt{3}$ | $\sqrt{14}$ | 1/2 |
| (1,1,3,2,1,4) | 6-18 | 2 | 504 | 6 | $2\sqrt{3}$ | $\sqrt{11}$ | $\sqrt{2}$ | 1/2 |
| (1,2,3,1,1,4) | 6-18 $I$ | 2 | 504 | 6 | $2\sqrt{3}$ | $\sqrt{11}$ | $\sqrt{2}$ | 1/2 |
| (1,1,2,1,3,4) | 6-15 | 2 | 0 | 5 | $2\sqrt{2}$ | $2\sqrt{3}$ | $2\sqrt{2}$ | $1/\sqrt{2}$ |
| (3,1,2,1,1,4) | 6-15 $I$ | 2 | 0 | 5 | $2\sqrt{2}$ | $2\sqrt{3}$ | $2\sqrt{2}$ | $1/\sqrt{2}$ |
| (2,1,1,1,3,4) | 6-Z39 | 2 | 0 | 5 | $\sqrt{5}$ | $2\sqrt{3}$ | $\sqrt{14}$ | 1/2 |
| (3,1,1,1,2,4) | 6-Z39 $I$ | 2 | 0 | 5 | $\sqrt{5}$ | $2\sqrt{3}$ | $\sqrt{14}$ | 1/2 |
| (1,1,3,1,2,4) | 6-Z43 | 2 | 0 | 5 | $\sqrt{11}$ | $2\sqrt{3}$ | $\sqrt{2}$ | 1/2 |
| (2,1,3,1,1,4) | 6-Z43 $I$ | 2 | 0 | 5 | $\sqrt{11}$ | $2\sqrt{3}$ | $\sqrt{2}$ | 1/2 |
| (1,1,2,2,2,4) | 6-22 | $\sqrt{3}$ | 1512 | 6 | $\sqrt{7}$ | $2\sqrt{2}$ | $\sqrt{6}$ | $\sqrt{2/3}$ |
| (2,2,2,1,1,4) | 6-22 $I$ | $\sqrt{3}$ | 1512 | 6 | $\sqrt{7}$ | $2\sqrt{2}$ | $\sqrt{6}$ | $\sqrt{2/3}$ |
| (1,2,2,1,2,4) | 6-Z25 | $\sqrt{3}$ | 1152 | 6 | $2\sqrt{2}$ | $2\sqrt{2}$ | 2 | $\sqrt{2/3}$ |
| (2,1,2,2,1,4) | 6-Z25 $I$ | $\sqrt{3}$ | 1152 | 6 | $2\sqrt{2}$ | $2\sqrt{2}$ | 2 | $\sqrt{2/3}$ |
| (1,1,3,1,3,3) | 6-Z44 | $\sqrt{3}$ | 1152 | 6 | $2\sqrt{2}$ | $2\sqrt{2}$ | 2 | $2/\sqrt{3}$ |
| (1,3,1,1,3,3) | 6-Z44 $I$ | $\sqrt{3}$ | 1152 | 6 | $2\sqrt{2}$ | $2\sqrt{2}$ | 2 | $2/\sqrt{3}$ |
| (1,2,1,2,2,4) | 6-Z24 | $\sqrt{3}$ | 1008 | 6 | $2\sqrt{2}$ | $\sqrt{5}$ | $\sqrt{10}$ | $1/\sqrt{3}$ |
| (2,2,1,2,1,4) | 6-Z24 $I$ | $\sqrt{3}$ | 1008 | 6 | $2\sqrt{2}$ | $\sqrt{5}$ | $\sqrt{10}$ | $1/\sqrt{3}$ |
| (2,1,1,2,2,4) | 6-21 | $\sqrt{3}$ | 936 | 6 | $\sqrt{5}$ | $2\sqrt{2}$ | $\sqrt{10}$ | $1/\sqrt{3}$ |
| (2,2,1,1,2,4) | 6-21 $I$ | $\sqrt{3}$ | 936 | 6 | $\sqrt{5}$ | $2\sqrt{2}$ | $\sqrt{10}$ | $1/\sqrt{3}$ |
| (1,2,2,2,1,4) | 6-Z26 | $\sqrt{3}$ | 432 | 6 | $\sqrt{10}$ | $\sqrt{5}$ | $\sqrt{6}$ | 0 |
| (2,1,2,1,2,4) | 6-Z23 | $\sqrt{3}$ | 0 | 5 | $\sqrt{6}$ | 3 | $\sqrt{6}$ | 0 |
| (1,1,1,3,3,3) | 6-Z42 | $\sqrt{3}$ | 0 | 4 | 2 | $2\sqrt{2}$ | $2\sqrt{3}$ | 0 |
| (1,3,1,3,1,3) | 6-20 | $\sqrt{3}$ | 0 | 2 | $2\sqrt{3}$ | 0 | $2\sqrt{3}$ | 0 |
| (1,2,1,2,3,3) | 6-27 | $\sqrt{2}$ | 504 | 6 | 2 | $\sqrt{5}$ | $\sqrt{6}$ | $1/\sqrt{2}$ |
| (2,1,2,1,3,3) | 6-27 $I$ | $\sqrt{2}$ | 504 | 6 | 2 | $\sqrt{5}$ | $\sqrt{6}$ | $1/\sqrt{2}$ |
| (1,1,2,3,2,3) | 6-Z47 | $\sqrt{2}$ | 504 | 6 | 2 | $\sqrt{5}$ | $\sqrt{6}$ | $1/\sqrt{2}$ |
| (2,1,1,3,2,3) | 6-Z47 $I$ | $\sqrt{2}$ | 504 | 6 | 2 | $\sqrt{5}$ | $\sqrt{6}$ | $1/\sqrt{2}$ |
| (2,1,3,1,2,3) | 6-Z29 | $\sqrt{2}$ | 216 | 6 | $\sqrt{6}$ | $\sqrt{5}$ | $\sqrt{2}$ | 0 |
| (1,3,2,1,2,3) | 6-Z50 | $\sqrt{2}$ | 216 | 6 | $\sqrt{6}$ | $\sqrt{5}$ | $\sqrt{2}$ | 0 |
| (1,3,1,2,2,3) | 6-31 | $\sqrt{2}$ | 144 | 6 | $\sqrt{7}$ | $\sqrt{2}$ | $\sqrt{6}$ | $1/\sqrt{2}$ |

| | | | | | | | | |
|---|---|---|---|---|---|---|---|---|
| (2,2,1,3,1,3) | 6-31 $I$ | $\sqrt{2}$ | 144 | 6 | $\sqrt{7}$ | $\sqrt{2}$ | $\sqrt{6}$ | $1/\sqrt{2}$ |
| (1,2,2,1,3,3) | 6-Z28 | $\sqrt{2}$ | 0 | 5 | $\sqrt{5}$ | $\sqrt{6}$ | $\sqrt{2}$ | 0 |
| (1,1,2,2,3,3) | 6-Z46 | $\sqrt{2}$ | 0 | 5 | $\sqrt{3}$ | $\sqrt{6}$ | $\sqrt{6}$ | 1 |
| (2,2,1,1,3,3) | 6-Z46 $I$ | $\sqrt{2}$ | 0 | 5 | $\sqrt{3}$ | $\sqrt{6}$ | $\sqrt{6}$ | 1 |
| (1,1,3,2,2,3) | 6-Z48 | $\sqrt{2}$ | 0 | 5 | $\sqrt{5}$ | $\sqrt{6}$ | $\sqrt{2}$ | 0 |
| (1,2,1,3,2,3) | 6-Z49 | $\sqrt{2}$ | 0 | 4 | $\sqrt{6}$ | $\sqrt{2}$ | $2\sqrt{2}$ | 0 |
| (1,2,3,1,2,3) | 6-30 | $\sqrt{2}$ | 0 | 3 | $\sqrt{6}$ | $\sqrt{6}$ | 0 | 1 |
| (2,1,3,2,1,3) | 6-30 $I$ | $\sqrt{2}$ | 0 | 3 | $\sqrt{6}$ | $\sqrt{6}$ | 0 | 1 |
| (2,1,1,2,3,3) | 6-Z45 | $\sqrt{2}$ | 0 | 3 | $\sqrt{2}$ | $\sqrt{6}$ | $2\sqrt{2}$ | 0 |
| (1,2,2,2,2,3) | 6-34 | 1 | 72 | 6 | $\sqrt{3}$ | $\sqrt{2}$ | $\sqrt{2}$ | 1 |
| (2,2,2,2,1,3) | 6-34 $I$ | 1 | 72 | 6 | $\sqrt{3}$ | $\sqrt{2}$ | $\sqrt{2}$ | 1 |
| (2,1,2,2,2,3) | 6-33 | 1 | 0 | 5 | $\sqrt{2}$ | $\sqrt{3}$ | $\sqrt{2}$ | 1 |
| (2,2,2,1,2,3) | 6-33 $I$ | 1 | 0 | 5 | $\sqrt{2}$ | $\sqrt{3}$ | $\sqrt{2}$ | 1 |
| (2,2,1,2,2,3) | 6-32 | 1 | 0 | 4 | $\sqrt{2}$ | $\sqrt{2}$ | 2 | 0 |
| (2,2,2,2,2,2) | 6-35 | 0 | 0 | 1 | 0 | 0 | 0 | 0 |

## $n = 7$

| Prime form | Forte Name | $\sqrt{7}\lVert P \rVert$ | $\lvert \det(M_H) \rvert$ | rank$(M_H)$ | $\widehat{\sigma_1}$ | $\widehat{\sigma_2}$ | $\widehat{\sigma_3}$ | $I$–sym |
|---|---|---|---|---|---|---|---|---|
| (1,1,1,1,1,1,6) | 7-01 | $5\sqrt{3}$ | 187500 | 7 | 5 | 5 | 5 | 0 |
| (1,1,1,1,1,2,5) | 7-02 | $\sqrt{47}$ | 39324 | 7 | $\sqrt{13}$ | $\sqrt{17}$ | $\sqrt{17}$ | $\sqrt{7/47}$ |
| (2,1,1,1,1,1,5) | 7-02 $I$ | $\sqrt{47}$ | 39324 | 7 | $\sqrt{13}$ | $\sqrt{17}$ | $\sqrt{17}$ | $\sqrt{7/47}$ |
| (1,1,1,1,2,1,5) | 7-04 | $\sqrt{47}$ | 39324 | 7 | $\sqrt{17}$ | $\sqrt{13}$ | $\sqrt{17}$ | $\sqrt{7/47}$ |
| (1,2,1,1,1,1,5) | 7-04 $I$ | $\sqrt{47}$ | 39324 | 7 | $\sqrt{17}$ | $\sqrt{13}$ | $\sqrt{17}$ | $\sqrt{7/47}$ |
| (1,1,1,2,1,1,5) | 7-05 | $\sqrt{47}$ | 39324 | 7 | $\sqrt{17}$ | $\sqrt{17}$ | $\sqrt{13}$ | $\sqrt{7/47}$ |
| (1,1,2,1,1,1,5) | 7-05 $I$ | $\sqrt{47}$ | 39324 | 7 | $\sqrt{17}$ | $\sqrt{17}$ | $\sqrt{13}$ | $\sqrt{7/47}$ |
| (1,1,1,1,1,3,4) | 7-03 | $\sqrt{33}$ | 5556 | 7 | $\sqrt{7}$ | $\sqrt{13}$ | $\sqrt{13}$ | $\sqrt{7/33}$ |
| (3,1,1,1,1,1,4) | 7-03 $I$ | $\sqrt{33}$ | 5556 | 7 | $\sqrt{7}$ | $\sqrt{13}$ | $\sqrt{13}$ | $\sqrt{7/33}$ |
| (1,1,1,1,3,1,4) | 7-06 | $\sqrt{33}$ | 5556 | 7 | $\sqrt{13}$ | $\sqrt{7}$ | $\sqrt{13}$ | $\sqrt{7/33}$ |
| (1,3,1,1,1,1,4) | 7-06 $I$ | $\sqrt{33}$ | 5556 | 7 | $\sqrt{13}$ | $\sqrt{7}$ | $\sqrt{13}$ | $\sqrt{7/33}$ |
| (1,1,1,3,1,1,4) | 7-07 | $\sqrt{33}$ | 5556 | 7 | $\sqrt{13}$ | $\sqrt{13}$ | $\sqrt{7}$ | $\sqrt{7/33}$ |
| (1,1,3,1,1,1,4) | 7-07 $I$ | $\sqrt{33}$ | 5556 | 7 | $\sqrt{13}$ | $\sqrt{13}$ | $\sqrt{7}$ | $\sqrt{7/33}$ |
| (1,2,1,1,1,2,4) | 7-11 | $\sqrt{26}$ | 6816 | 7 | $2\sqrt{2}$ | $2\sqrt{2}$ | $\sqrt{10}$ | $\sqrt{7/13}$ |
| (2,1,1,1,2,1,4) | 7-11 $I$ | $\sqrt{26}$ | 6816 | 7 | $2\sqrt{2}$ | $2\sqrt{2}$ | $\sqrt{10}$ | $\sqrt{7/13}$ |
| (1,1,1,2,1,2,4) | 7-Z36 | $\sqrt{26}$ | 6816 | 7 | $2\sqrt{2}$ | $\sqrt{10}$ | $2\sqrt{2}$ | $\sqrt{7/13}$ |
| (2,1,2,1,1,1,4) | 7-Z36 $I$ | $\sqrt{26}$ | 6816 | 7 | $2\sqrt{2}$ | $\sqrt{10}$ | $2\sqrt{2}$ | $\sqrt{7/13}$ |
| (1,1,1,2,2,1,4) | 7-14 | $\sqrt{26}$ | 6816 | 7 | $\sqrt{10}$ | $2\sqrt{2}$ | $2\sqrt{2}$ | $\sqrt{7/13}$ |
| (1,2,2,1,1,1,4) | 7-14 $I$ | $\sqrt{26}$ | 6816 | 7 | $\sqrt{10}$ | $2\sqrt{2}$ | $2\sqrt{2}$ | $\sqrt{7/13}$ |
| (1,1,1,1,2,2,4) | 7-09 | $\sqrt{26}$ | 5052 | 7 | $\sqrt{7}$ | $2\sqrt{2}$ | $\sqrt{11}$ | $\sqrt{7/13}$ |
| (2,2,1,1,1,1,4) | 7-09 $I$ | $\sqrt{26}$ | 5052 | 7 | $\sqrt{7}$ | $2\sqrt{2}$ | $\sqrt{11}$ | $\sqrt{7/13}$ |
| (1,1,2,1,1,2,4) | 7-13 | $\sqrt{26}$ | 5052 | 7 | $2\sqrt{2}$ | $\sqrt{11}$ | $\sqrt{7}$ | $\sqrt{7/13}$ |
| (2,1,1,2,1,1,4) | 7-13 $I$ | $\sqrt{26}$ | 5052 | 7 | $2\sqrt{2}$ | $\sqrt{11}$ | $\sqrt{7}$ | $\sqrt{7/13}$ |
| (1,1,2,1,2,1,4) | 7-Z38 | $\sqrt{26}$ | 5052 | 7 | $\sqrt{11}$ | $\sqrt{7}$ | $2\sqrt{2}$ | $\sqrt{7/13}$ |
| (1,2,1,2,1,1,4) | 7-Z38 $I$ | $\sqrt{26}$ | 5052 | 7 | $\sqrt{11}$ | $\sqrt{7}$ | $2\sqrt{2}$ | $\sqrt{7/13}$ |
| (2,1,1,1,1,2,4) | 7-08 | $\sqrt{26}$ | 2028 | 7 | $\sqrt{5}$ | $\sqrt{10}$ | $\sqrt{11}$ | 0 |

| | | | | | | | | |
|---|---|---|---|---|---|---|---|---|
| (1,1,2,2,1,1,4) | 7-15 | $\sqrt{26}$ | 2028 | 7 | $\sqrt{10}$ | $\sqrt{11}$ | $\sqrt{5}$ | 0 |
| (1,2,1,1,2,1,4) | 7-Z37 | $\sqrt{26}$ | 2028 | 7 | $\sqrt{11}$ | $\sqrt{5}$ | $\sqrt{10}$ | 0 |
| (1,1,1,2,1,3,3) | 7-16 | $\sqrt{19}$ | 2364 | 7 | $\sqrt{5}$ | $\sqrt{7}$ | $\sqrt{7}$ | $\sqrt{7/19}$ |
| (1,2,1,1,1,3,3) | 7-16 $I$ | $\sqrt{19}$ | 2364 | 7 | $\sqrt{5}$ | $\sqrt{7}$ | $\sqrt{7}$ | $\sqrt{7/19}$ |
| (1,3,1,1,1,2,3) | 7-Z18 | $\sqrt{19}$ | 2364 | 7 | $\sqrt{7}$ | $\sqrt{5}$ | $\sqrt{7}$ | $\sqrt{7/19}$ |
| (2,1,1,1,3,1,3) | 7-Z18 $I$ | $\sqrt{19}$ | 2364 | 7 | $\sqrt{7}$ | $\sqrt{5}$ | $\sqrt{7}$ | $\sqrt{7/19}$ |
| (1,1,1,3,1,2,3) | 7-19 | $\sqrt{19}$ | 2364 | 7 | $\sqrt{7}$ | $\sqrt{7}$ | $\sqrt{5}$ | $\sqrt{7/19}$ |
| (1,1,1,3,2,1,3) | 7-19 $I$ | $\sqrt{19}$ | 2364 | 7 | $\sqrt{7}$ | $\sqrt{7}$ | $\sqrt{5}$ | $\sqrt{7/19}$ |
| (1,1,1,1,2,3,3) | 7-10 | $\sqrt{19}$ | 348 | 7 | $\sqrt{3}$ | $\sqrt{7}$ | 3 | $\sqrt{7/19}$ |
| (2,1,1,1,1,3,3) | 7-10 $I$ | $\sqrt{19}$ | 348 | 7 | $\sqrt{3}$ | $\sqrt{7}$ | 3 | $\sqrt{7/19}$ |
| (1,1,3,1,1,2,3) | 7-20 | $\sqrt{19}$ | 348 | 7 | $\sqrt{7}$ | 3 | $\sqrt{3}$ | $\sqrt{7/19}$ |
| (2,1,1,3,1,1,3) | 7-20 $I$ | $\sqrt{19}$ | 348 | 7 | $\sqrt{7}$ | 3 | $\sqrt{3}$ | $\sqrt{7/19}$ |
| (1,1,2,1,3,1,3) | 7-21 | $\sqrt{19}$ | 348 | 7 | 3 | $\sqrt{3}$ | $\sqrt{7}$ | $\sqrt{7/19}$ |
| (1,2,1,1,3,1,3) | 7-21 $I$ | $\sqrt{19}$ | 348 | 7 | 3 | $\sqrt{3}$ | $\sqrt{7}$ | $\sqrt{7/19}$ |
| (1,1,2,1,1,3,3) | 7-Z17 | $\sqrt{19}$ | 12 | 7 | $\sqrt{5}$ | 9 | $\sqrt{5}$ | 0 |
| (1,1,1,1,3,2,3) | 7-Z12 | $\sqrt{19}$ | 12 | 7 | $\sqrt{5}$ | $\sqrt{5}$ | 3 | 0 |
| (1,1,3,1,2,1,3) | 7-22 | $\sqrt{19}$ | 12 | 7 | 9 | $\sqrt{5}$ | $\sqrt{5}$ | 0 |
| (1,1,2,1,2,2,3) | 7-27 | $2\sqrt{3}$ | 768 | 7 | 2 | 2 | 2 | $\sqrt{7/6}$ |
| (2,2,1,2,1,1,3) | 7-27 $I$ | $2\sqrt{3}$ | 768 | 7 | 2 | 2 | 2 | $\sqrt{7/6}$ |
| (1,1,1,2,2,2,3) | 7-24 | $2\sqrt{3}$ | 516 | 7 | $\sqrt{3}$ | 2 | $\sqrt{5}$ | $\sqrt{7}/(2\sqrt{3})$ |
| (2,2,2,1,1,1,3) | 7-24 $I$ | $2\sqrt{3}$ | 516 | 7 | $\sqrt{3}$ | 2 | $\sqrt{5}$ | $\sqrt{7}/(2\sqrt{3})$ |
| (1,2,2,1,1,2,3) | 7-28 | $2\sqrt{3}$ | 516 | 7 | 2 | $\sqrt{5}$ | $\sqrt{3}$ | $\sqrt{7}/(2\sqrt{3})$ |
| (2,1,1,2,2,1,3) | 7-28 $I$ | $2\sqrt{3}$ | 516 | 7 | 2 | $\sqrt{5}$ | $\sqrt{3}$ | $\sqrt{7}/(2\sqrt{3})$ |
| (1,2,1,2,1,2,3) | 7-31 | $2\sqrt{3}$ | 516 | 7 | $\sqrt{5}$ | $\sqrt{3}$ | 2 | $\sqrt{7}/(2\sqrt{3})$ |
| (2,1,2,1,2,1,3) | 7-31 $I$ | $2\sqrt{3}$ | 516 | 7 | $\sqrt{5}$ | $\sqrt{3}$ | 2 | $\sqrt{7}/(2\sqrt{3})$ |
| (2,1,1,2,1,2,3) | 7-25 | $2\sqrt{3}$ | 348 | 7 | $\sqrt{3}$ | $\sqrt{5}$ | 2 | $\sqrt{7}/(2\sqrt{3})$ |
| (2,1,2,1,1,2,3) | 7-25 $I$ | $2\sqrt{3}$ | 348 | 7 | $\sqrt{3}$ | $\sqrt{5}$ | 2 | $\sqrt{7}/(2\sqrt{3})$ |
| (1,2,1,1,2,2,3) | 7-26 | $2\sqrt{3}$ | 348 | 7 | 2 | $\sqrt{3}$ | $\sqrt{5}$ | $\sqrt{7}/(2\sqrt{3})$ |
| (2,2,1,1,2,1,3) | 7-26 $I$ | $2\sqrt{3}$ | 348 | 7 | 2 | $\sqrt{3}$ | $\sqrt{5}$ | $\sqrt{7}/(2\sqrt{3})$ |
| (1,1,2,2,2,1,3) | 7-30 | $2\sqrt{3}$ | 348 | 7 | $\sqrt{5}$ | 2 | $\sqrt{3}$ | $\sqrt{7}/(2\sqrt{3})$ |
| (1,2,2,2,1,1,3) | 7-30 $I$ | $2\sqrt{3}$ | 348 | 7 | $\sqrt{5}$ | 2 | $\sqrt{3}$ | $\sqrt{7}/(2\sqrt{3})$ |
| (2,1,1,1,2,2,3) | 7-23 | $2\sqrt{3}$ | 96 | 7 | $\sqrt{2}$ | 2 | $\sqrt{6}$ | $\sqrt{7}/(2\sqrt{3})$ |
| (2,2,1,1,1,2,3) | 7-23 $I$ | $2\sqrt{3}$ | 96 | 7 | $\sqrt{2}$ | 2 | $\sqrt{6}$ | $\sqrt{7}/(2\sqrt{3})$ |
| (1,1,2,2,1,2,3) | 7-29 | $2\sqrt{3}$ | 96 | 7 | 2 | $\sqrt{6}$ | $\sqrt{2}$ | $\sqrt{7}/(2\sqrt{3})$ |
| (2,1,2,2,1,1,3) | 7-29 $I$ | $2\sqrt{3}$ | 96 | 7 | 2 | $\sqrt{6}$ | $\sqrt{2}$ | $\sqrt{7}/(2\sqrt{3})$ |
| (1,2,1,2,2,1,3) | 7-32 | $2\sqrt{3}$ | 96 | 7 | $\sqrt{6}$ | $\sqrt{2}$ | 2 | $\sqrt{7}/(2\sqrt{3})$ |
| (1,2,2,1,2,1,3) | 7-32 $I$ | $2\sqrt{3}$ | 96 | 7 | $\sqrt{6}$ | $\sqrt{2}$ | 2 | $\sqrt{7}/(2\sqrt{3})$ |
| (1,1,2,2,2,2,2) | 7-33 | $\sqrt{5}$ | 12 | 7 | 1 | $\sqrt{2}$ | $\sqrt{2}$ | 0 |
| (1,2,1,2,2,2,2) | 7-34 | $\sqrt{5}$ | 12 | 7 | $\sqrt{2}$ | 1 | $\sqrt{2}$ | 0 |
| (1,2,2,1,2,2,2) | 7-35 | $\sqrt{5}$ | 12 | 7 | $\sqrt{2}$ | $\sqrt{2}$ | 1 | 0 |

**n = 8**

| Prime form | Forte Name | Index $\|P\|$ | $\det(M_H)$ | rank$(M_H)$ | $\widehat{\sigma_1}$ | $\widehat{\sigma_2}$ | $\widehat{\sigma_3}$ | $\widehat{\sigma_4}$ | I–sym |
|---|---|---|---|---|---|---|---|---|---|
| (1,1,1,1,1,1,1,5) | 8-01 | $\sqrt{7}$ | 196608 | 8 | 4 | 4 | 4 | 4 | 0 |
| (1,1,1,1,1,1,2,4) | 8-02 | 2 | 19680 | 8 | $\sqrt{7}$ | $\sqrt{10}$ | $\sqrt{10}$ | $\sqrt{10}$ | 1/2 |
| (2,1,1,1,1,1,1,4) | 8-02 I | 2 | 19680 | 8 | $\sqrt{7}$ | $\sqrt{10}$ | $\sqrt{10}$ | $\sqrt{10}$ | 1/2 |
| (1,1,1,1,2,1,1,4) | 8-05 | 2 | 19680 | 8 | $\sqrt{10}$ | $\sqrt{10}$ | $\sqrt{7}$ | $\sqrt{10}$ | 1/2 |
| (1,1,2,1,1,1,1,4) | 8-05 I | 2 | 19680 | 8 | $\sqrt{10}$ | $\sqrt{10}$ | $\sqrt{7}$ | $\sqrt{10}$ | 1/2 |
| (1,1,1,1,1,2,1,4) | 8-04 | 2 | 19200 | 8 | $\sqrt{10}$ | $\sqrt{7}$ | $\sqrt{10}$ | $\sqrt{10}$ | 1/2 |
| (1,2,1,1,1,1,1,4) | 8-04 I | 2 | 19200 | 8 | $\sqrt{10}$ | $\sqrt{7}$ | $\sqrt{10}$ | $\sqrt{10}$ | 1/2 |
| (1,1,1,2,1,1,1,4) | 8-06 | 2 | 12288 | 8 | $\sqrt{10}$ | $\sqrt{10}$ | $\sqrt{10}$ | 2 | 0 |
| (1,1,1,1,1,1,3,3) | 8-03 | $\sqrt{3}$ | 0 | 7 | 2 | $2\sqrt{2}$ | $2\sqrt{2}$ | $2\sqrt{2}$ | 0 |
| (1,1,1,1,3,1,1,3) | 8-08 | $\sqrt{3}$ | 0 | 7 | $2\sqrt{2}$ | $2\sqrt{2}$ | 2 | $2\sqrt{2}$ | 0 |
| (1,1,1,1,1,3,1,3) | 8-07 | $\sqrt{3}$ | 0 | 6 | $2\sqrt{2}$ | 2 | $2\sqrt{2}$ | $2\sqrt{2}$ | 0 |
| (1,1,1,3,1,1,1,3) | 8-09 | $\sqrt{3}$ | 0 | 4 | $2\sqrt{2}$ | $2\sqrt{2}$ | $2\sqrt{2}$ | 0 | 0 |
| (1,2,1,1,1,1,2,3) | 8-12 | $\sqrt{2}$ | 1632 | 8 | 2 | 2 | $\sqrt{5}$ | $\sqrt{6}$ | $\sqrt{2}$ |
| (2,1,1,1,1,2,1,3) | 8-12 I | $\sqrt{2}$ | 1632 | 8 | 2 | 2 | $\sqrt{5}$ | $\sqrt{6}$ | $\sqrt{2}$ |
| (1,1,1,1,2,2,1,3) | 8-Z15 | $\sqrt{2}$ | 1632 | 8 | $\sqrt{5}$ | 2 | 2 | $\sqrt{6}$ | $\sqrt{2}$ |
| (1,2,2,1,1,1,1,3) | 8-Z15 I | $\sqrt{2}$ | 1632 | 8 | $\sqrt{5}$ | 2 | 2 | $\sqrt{6}$ | $\sqrt{2}$ |
| (1,1,1,1,1,2,2,3) | 8-11 | $\sqrt{2}$ | 864 | 8 | $\sqrt{3}$ | 2 | $\sqrt{6}$ | $\sqrt{6}$ | $1/\sqrt{2}$ |
| (2,2,1,1,1,1,1,3) | 8-11 I | $\sqrt{2}$ | 864 | 8 | $\sqrt{3}$ | 2 | $\sqrt{6}$ | $\sqrt{6}$ | $1/\sqrt{2}$ |
| (1,1,2,1,1,2,1,3) | 8-19 | $\sqrt{2}$ | 864 | 8 | $\sqrt{6}$ | 2 | $\sqrt{3}$ | $\sqrt{6}$ | $1/\sqrt{2}$ |
| (1,2,1,1,2,1,1,3) | 8-19 I | $\sqrt{2}$ | 864 | 8 | $\sqrt{6}$ | 2 | $\sqrt{3}$ | $\sqrt{6}$ | $1/\sqrt{2}$ |
| (1,1,1,2,1,2,1,3) | 8-18 | $\sqrt{2}$ | 768 | 8 | $\sqrt{6}$ | $\sqrt{3}$ | $\sqrt{6}$ | $\sqrt{2}$ | $1/\sqrt{2}$ |
| (1,2,1,2,1,1,1,3) | 8-18 I | $\sqrt{2}$ | 768 | 8 | $\sqrt{6}$ | $\sqrt{3}$ | $\sqrt{6}$ | $\sqrt{2}$ | $1/\sqrt{2}$ |
| (1,1,1,2,2,1,1,3) | 8-16 | $\sqrt{2}$ | 480 | 8 | $\sqrt{5}$ | $\sqrt{6}$ | 2 | $\sqrt{2}$ | $1/\sqrt{2}$ |
| (1,1,2,2,1,1,1,3) | 8-16 I | $\sqrt{2}$ | 480 | 8 | $\sqrt{5}$ | $\sqrt{6}$ | 2 | $\sqrt{2}$ | $1/\sqrt{2}$ |
| (1,1,1,2,1,1,2,3) | 8-Z29 | $\sqrt{2}$ | 480 | 8 | 2 | $\sqrt{6}$ | $\sqrt{5}$ | $\sqrt{2}$ | $1/\sqrt{2}$ |
| (2,1,1,2,1,1,1,3) | 8-Z29 I | $\sqrt{2}$ | 480 | 8 | 2 | $\sqrt{6}$ | $\sqrt{5}$ | $\sqrt{2}$ | $1/\sqrt{2}$ |
| (2,1,1,1,1,1,2,3) | 8-10 | $\sqrt{2}$ | 0 | 7 | $\sqrt{2}$ | $\sqrt{5}$ | $\sqrt{6}$ | $\sqrt{6}$ | 0 |
| (1,1,1,1,2,1,2,3) | 8-13 | $\sqrt{2}$ | 0 | 7 | 2 | $\sqrt{5}$ | 2 | $\sqrt{6}$ | $\sqrt{2}$ |
| (2,1,2,1,1,1,1,3) | 8-13 I | $\sqrt{2}$ | 0 | 7 | 2 | $\sqrt{5}$ | 2 | $\sqrt{6}$ | $\sqrt{2}$ |
| (1,1,2,1,1,1,2,3) | 8-14 | $\sqrt{2}$ | 0 | 7 | 2 | $\sqrt{6}$ | 2 | 2 | $\sqrt{2}$ |
| (2,1,1,1,2,1,1,3) | 8-14 I | $\sqrt{2}$ | 0 | 7 | 2 | $\sqrt{6}$ | 2 | 2 | $\sqrt{2}$ |
| (1,1,2,1,2,1,1,3) | 8-20 | $\sqrt{2}$ | 0 | 7 | $\sqrt{6}$ | $\sqrt{5}$ | $\sqrt{2}$ | $\sqrt{6}$ | 0 |
| (1,2,1,1,1,2,1,3) | 8-17 | $\sqrt{2}$ | 0 | 6 | $\sqrt{6}$ | $\sqrt{2}$ | $\sqrt{6}$ | $\sqrt{2}$ | 0 |
| (1,1,1,2,1,2,2,2) | 8-22 | 1 | 96 | 8 | $\sqrt{2}$ | $\sqrt{2}$ | $\sqrt{3}$ | $\sqrt{2}$ | 1 |
| (1,2,1,1,1,2,2,2) | 8-22 I | 1 | 96 | 8 | $\sqrt{2}$ | $\sqrt{2}$ | $\sqrt{3}$ | $\sqrt{2}$ | 1 |
| (1,1,2,1,2,1,2,2) | 8-27 | 1 | 96 | 8 | $\sqrt{3}$ | $\sqrt{2}$ | $\sqrt{2}$ | $\sqrt{2}$ | 1 |
| (1,2,1,2,1,1,2,2) | 8-27 I | 1 | 96 | 8 | $\sqrt{3}$ | $\sqrt{2}$ | $\sqrt{2}$ | $\sqrt{2}$ | 1 |
| (1,1,2,1,1,2,2,2) | 8-24 | 1 | 0 | 7 | $\sqrt{2}$ | $\sqrt{3}$ | $\sqrt{2}$ | $\sqrt{2}$ | 0 |
| (1,1,1,2,2,1,2,2) | 8-23 | 1 | 0 | 7 | $\sqrt{2}$ | $\sqrt{3}$ | $\sqrt{2}$ | $\sqrt{2}$ | 0 |
| (1,1,1,1,2,2,2,2) | 8-21 | 1 | 0 | 5 | 1 | $\sqrt{2}$ | $\sqrt{3}$ | 2 | 0 |
| (1,2,1,1,2,1,2,2) | 8-26 | 1 | 0 | 5 | $\sqrt{3}$ | $\sqrt{2}$ | 1 | 2 | 0 |
| (1,1,2,2,1,1,2,2) | 8-25 | 1 | 0 | 3 | $\sqrt{2}$ | 2 | $\sqrt{2}$ | 0 | 0 |
| (1,2,1,2,1,2,1,2) | 8-28 | 1 | 0 | 2 | 2 | 0 | 2 | 0 | 0 |

**n = 9**

| Prime form | Forte Name | Index $\|P\|$ | $\|\det(M_H)\|$ | rank$(M_H)$ | $\widehat{\sigma_1}$ | $\widehat{\sigma_2}$ | $\widehat{\sigma_3}$ | $\widehat{\sigma_4}$ | $I$–sym |
|---|---|---|---|---|---|---|---|---|---|
| (1,1,1,1,1,1,1,1,4) | 9–01 | $\sqrt{2}$ | 78732 | 9 | 3 | 3 | 3 | 3 | 0 |
| (1,1,1,1,1,2,1,1,3) | 9–04 | $\sqrt{2}$ | 2916 | 9 | $\sqrt{5}$ | $\sqrt{5}$ | $\sqrt{3}$ | $\sqrt{5}$ | $1/\sqrt{2}$ |
| (1,1,2,1,1,1,1,1,3) | 9–04 $I$ | $\sqrt{2}$ | 2916 | 9 | $\sqrt{5}$ | $\sqrt{5}$ | $\sqrt{3}$ | $\sqrt{5}$ | $1/\sqrt{2}$ |
| (1,1,1,1,1,1,1,2,3) | 9–02 | $\sqrt{2}$ | 2052 | 9 | $\sqrt{3}$ | $\sqrt{5}$ | $\sqrt{5}$ | $\sqrt{5}$ | $1/\sqrt{2}$ |
| (2,1,1,1,1,1,1,1,3) | 9–02 $I$ | $\sqrt{2}$ | 2052 | 9 | $\sqrt{3}$ | $\sqrt{5}$ | $\sqrt{5}$ | $\sqrt{5}$ | $1/\sqrt{2}$ |
| (1,1,1,1,1,1,2,1,3) | 9–03 | $\sqrt{2}$ | 2052 | 9 | $\sqrt{5}$ | $\sqrt{3}$ | $\sqrt{5}$ | $\sqrt{5}$ | $1/\sqrt{2}$ |
| (1,2,1,1,1,1,1,1,3) | 9–03 $I$ | $\sqrt{2}$ | 2052 | 9 | $\sqrt{5}$ | $\sqrt{3}$ | $\sqrt{5}$ | $\sqrt{5}$ | $1/\sqrt{2}$ |
| (1,1,1,1,2,1,1,1,3) | 9–05 | $\sqrt{2}$ | 2052 | 9 | $\sqrt{5}$ | $\sqrt{5}$ | $\sqrt{5}$ | $\sqrt{3}$ | $1/\sqrt{2}$ |
| (1,1,1,2,1,1,1,1,3) | 9–05 $I$ | $\sqrt{2}$ | 2052 | 9 | $\sqrt{5}$ | $\sqrt{5}$ | $\sqrt{5}$ | $\sqrt{3}$ | $1/\sqrt{2}$ |
| (1,1,1,1,1,2,1,2,2) | 9–07 | 1 | 108 | 9 | $\sqrt{2}$ | $\sqrt{2}$ | $\sqrt{2}$ | $\sqrt{3}$ | 1 |
| (1,2,1,1,1,1,1,2,2) | 9–07 $I$ | 1 | 108 | 9 | $\sqrt{2}$ | $\sqrt{2}$ | $\sqrt{2}$ | $\sqrt{3}$ | 1 |
| (1,1,1,1,2,1,1,2,2) | 9–08 | 1 | 108 | 9 | $\sqrt{2}$ | $\sqrt{3}$ | $\sqrt{2}$ | $\sqrt{2}$ | 1 |
| (1,1,2,1,1,1,1,2,2) | 9–08 $I$ | 1 | 108 | 9 | $\sqrt{2}$ | $\sqrt{3}$ | $\sqrt{2}$ | $\sqrt{2}$ | 1 |
| (1,1,1,2,1,1,2,1,2) | 9–11 | 1 | 108 | 9 | $\sqrt{3}$ | $\sqrt{2}$ | $\sqrt{2}$ | $\sqrt{2}$ | 1 |
| (1,1,2,1,1,1,2,1,2) | 9–11 $I$ | 1 | 108 | 9 | $\sqrt{3}$ | $\sqrt{2}$ | $\sqrt{2}$ | $\sqrt{2}$ | 1 |
| (1,1,1,1,1,1,2,2,2) | 9–06 | 1 | 0 | 7 | 1 | $\sqrt{2}$ | $\sqrt{3}$ | $\sqrt{3}$ | 0 |
| (1,1,1,2,1,1,1,2,2) | 9–09 | 1 | 0 | 7 | $\sqrt{2}$ | $\sqrt{3}$ | $\sqrt{3}$ | 1 | 0 |
| (1,1,1,1,2,1,2,1,2) | 9–10 | 1 | 0 | 7 | $\sqrt{3}$ | 1 | $\sqrt{3}$ | $\sqrt{2}$ | 0 |
| (1,1,2,1,1,2,1,1,2) | 9–12 | 1 | 0 | 3 | $\sqrt{3}$ | $\sqrt{3}$ | 0 | $\sqrt{3}$ | 0 |

**n = 10**

| Prime form | Forte Name | Index $\|P\|$ | $\|\det(M_H)\|$ | rank$(M_H)$ | $\widehat{\sigma_1}$ | $\widehat{\sigma_2}$ | $\widehat{\sigma_3}$ | $\widehat{\sigma_4}$ | $\widehat{\sigma_5}$ | $I$–sym |
|---|---|---|---|---|---|---|---|---|---|---|
| (1,1,1,1,1,1,1,1,1,3) | 10–01 | $3/\sqrt{5}$ | 6144 | 10 | 2 | 2 | 2 | 2 | 2 | 0 |
| (1,1,1,1,1,1,1,2,1,2) | 10–03 | $2/\sqrt{5}$ | 24 | 10 | $\sqrt{2}$ | 1 | $\sqrt{2}$ | $\sqrt{2}$ | $\sqrt{2}$ | 0 |
| (1,1,1,1,1,2,1,1,1,2) | 10–05 | $2/\sqrt{5}$ | 24 | 10 | $\sqrt{2}$ | $\sqrt{2}$ | $\sqrt{2}$ | 1 | $\sqrt{2}$ | 0 |
| (1,1,1,1,1,1,1,1,2,2) | 10–02 | $2/\sqrt{5}$ | 0 | 9 | 1 | $\sqrt{2}$ | $\sqrt{2}$ | $\sqrt{2}$ | $\sqrt{2}$ | 0 |
| (1,1,1,1,1,1,2,1,1,2) | 10–04 | $2/\sqrt{5}$ | 0 | 9 | $\sqrt{2}$ | $\sqrt{2}$ | 1 | $\sqrt{2}$ | $\sqrt{2}$ | 0 |
| (1,1,1,1,2,1,1,1,1,2) | 10–06 | $2/\sqrt{5}$ | 0 | 5 | $\sqrt{2}$ | $\sqrt{2}$ | $\sqrt{2}$ | $\sqrt{2}$ | 0 | 0 |

**n = 11**

| Prime form | Index $\|P\|$ | $\|\det(M_H)\|$ | rank$(M_H)$ | $\widehat{\sigma_1}$ | $\widehat{\sigma_2}$ | $\widehat{\sigma_3}$ | $\widehat{\sigma_4}$ | $\widehat{\sigma_5}$ | $I$–sym |
|---|---|---|---|---|---|---|---|---|---|
| (1,1,1,1,1,1,1,1,1,1,2) | $\sqrt{5/11}$ | 12 | 11 | 1 | 1 | 1 | 1 | 1 | 0 |

## Appendix B. Dissonance of a chord and the volume content of its simplicial representation

We strongly emphasized that, in our model, two chords in $\mathcal{T}^n$ inscribed in the same $\mathbb{S}^{n-2}$ sphere have the same level of dissonance, because the dissonance is quantified as the $\delta$-radius of this sphere. However it is natural to ask: Is there any way to refine the dissonance comparisons amongst chords contained in the same $\mathbb{S}^{n-2}$ sphere, particularly in higher dimensions where there appears to be a significant number of such cases?

If we observe the global picture of our model, it seems plausible to ascertain that greater volume in our simplices corresponds to greater dissonance. Moreover, chords with more clustering distribution of intervals, admittedly more dissonant, bear a greater volume.

In view of these concerns, it appears reasonable to sequence the dissonance of chords that are inscribed in the same $\mathbb{S}^{n-2}$ by their increasing order of volumes. Once again we should emphasize that this ordering of dissonance by volume only applies to chords circumscribed with the same dissonance sphere, i.e. chords with the same $\delta$-radius of $\|P\|$.

We are going to derive a formula to calculate the volume of our simplicial chords in terms of the determinants of their Hankel matrices.

We will start from case $\mathcal{T}^3$ and extend the results to $\mathcal{T}^n$ in a natural manner.

Consider a chord $P = (x_1, x_2, x_3)$ in $\mathcal{T}^3$, then this chord and its orbit are immersed in the Euclidian space $\mathbb{R}^3$, i.e. with corresponding rectangular coordinates $(x, y, z) = (x_1, x_2, x_3)$ in $\mathbb{R}^3$. Assume furthermore, the coordinates $P$, $\sigma(P)$, $\sigma^2(P)$ to be rectangular. Let $h$ be the Euclidian distance between the origin in $\mathbb{R}^3$ and the barycenter of $\mathcal{T}^3$, $(x, y, z) = (4, 4, 4)$. (See Figure F.1.)

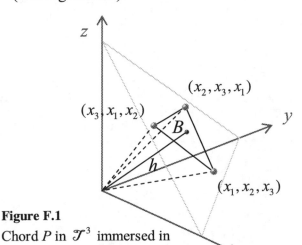

**Figure F.1**
Chord $P$ in $\mathcal{T}^3$ immersed in Euclidian space $\mathbb{R}^3$.

We know from Euclidian geometry that the volume of the pyramid with vertices at the origin and $P$, $\sigma(P)$, $\sigma^2(P)$ is:

$$V = \frac{1}{3!} \left| \det \begin{pmatrix} x_1 & x_2 & x_3 \\ x_2 & x_3 & x_1 \\ x_3 & x_1 & x_2 \end{pmatrix} \right| \quad \text{in } \mathbb{R}^3$$

This pyramid in $\mathbb{R}^3$ has an apex at the origin, and a corresponding base consisting of a triangle determined by $P$, $\sigma(P)$, $\sigma^2(P)$, with an altitude of $h$. Hence the triangular base area can be found to be:

$$A = \frac{3 \cdot V}{h} = \frac{\left| \det \begin{pmatrix} x_1 & x_2 & x_3 \\ x_2 & x_3 & x_1 \\ x_3 & x_1 & x_2 \end{pmatrix} \right|}{2! \, h} \quad \text{in } \mathbb{R}^3$$

Since the barycentric metric is $\dfrac{1}{\sqrt{2}}$ the Euclidian metric, the area measured in the barycentric coordinate system has an adjustment factor of $\dfrac{1}{\left(\sqrt{2}\right)^2}$ and thus the above area of the triangle in the barycentric model is:

$$A = \frac{\left| \det \begin{pmatrix} x_1 & x_2 & x_3 \\ x_2 & x_3 & x_1 \\ x_3 & x_1 & x_2 \end{pmatrix} \right|}{2! \left(\sqrt{2}\right)^2 \cdot h} \quad \text{in } \mathcal{T}^3$$

It is well known (Sommerville [8]) that the volume of an $n$-simplex with one vertex at the origin and the other $n$ vertices at $P = (x_1, x_2, x_3, \cdots, x_n)$, $\sigma(P) = (x_2, x_3, \cdots, x_n, x_1)$, $\sigma^{n-1}(P) = (x_n, x_1, x_2, \cdots, x_{n-1})$ in rectangular coordinates is

$$V_{n-simplex} = \frac{1}{n!} \left| \det \begin{pmatrix} x_1 & x_2 & x_3 & \cdots & x_n \\ x_2 & x_3 & x_4 & \cdots & x_1 \\ x_3 & x_4 & x_5 & \cdots & x_2 \\ \vdots & \vdots & \vdots & \ddots & \vdots \\ x_n & x_1 & x_2 & \cdots & x_{n-1} \end{pmatrix} \right| \quad \text{in } \mathbb{R}^n$$

The volume of an $n$-dimensional simplex is related to the $(n-1)$-dimensional volume of its base and the corresponding height $h$ as: $V_{n-simplex} = \dfrac{1}{n} V_{base} \cdot h$ in $\mathbb{R}^n$.

The volume of the $n$-chord $P$ corresponds to the $(n-1)$-dimensional base of the $n$-simplex, hence the volume of chord $P$ can be found to be: $V_{Chord} = \dfrac{n \cdot V_{n-simplex}}{h}$ in $\mathbb{R}^n$.

So the volume of the chord $P$ in $\mathcal{T}^n$ after the barycentric coordinate correction would be:

$$V(P) = \frac{\left| \det \begin{pmatrix} x_1 & x_2 & x_3 & \cdots & x_n \\ x_2 & x_3 & x_4 & \cdots & x_1 \\ x_3 & x_4 & x_5 & \cdots & x_2 \\ \vdots & \vdots & \vdots & \ddots & \vdots \\ x_n & x_1 & x_2 & \cdots & x_{n-1} \end{pmatrix} \right|}{(n-1)!\left(\sqrt{2}\right)^{n-1} \cdot h} \qquad \text{in } \mathcal{T}^n \qquad \text{(F.1)}$$

where $h$ is the Euclidian distance between the origin in $\mathbb{R}^n$ and the barycenter of $\mathcal{T}^n$.

Another interesting way of finding the volume of a simplex is in terms of its edge lengths via the Cayley-Menger determinant.

In our mesh notation the Cayley-Menger matrix of a chord $P$ in $\mathcal{T}^n$ will be:

$$\widehat{B}(P) = \begin{pmatrix} 0 & 1 & 1 & 1 & \cdots & 1 & 1 & \cdots & 1 \\ 1 & 0 & \widehat{\sigma}_1^{\,2} & \widehat{\sigma}_2^{\,2} & \cdots & \widehat{\sigma}_k^{\,2} & \widehat{\sigma}_{k-1}^{\,2} & \cdots & \widehat{\sigma}_1^{\,2} \\ 1 & \widehat{\sigma}_1^{\,2} & 0 & \widehat{\sigma}_1^{\,2} & \cdots & \widehat{\sigma}_{k-1}^{\,2} & \widehat{\sigma}_k^{\,2} & \cdots & \widehat{\sigma}_2^{\,2} \\ 1 & \widehat{\sigma}_2^{\,2} & \widehat{\sigma}_1^{\,2} & 0 & \widehat{\sigma}_1^{\,2} & \cdots & & & \vdots \\ \vdots & \vdots & \vdots & \widehat{\sigma}_1^{\,2} & 0 & \widehat{\sigma}_1^{\,2} & \cdots & & \widehat{\sigma}_k^{\,2} \\ 1 & \widehat{\sigma}_k^{\,2} & \widehat{\sigma}_{k-1}^{\,2} & \vdots & \widehat{\sigma}_1^{\,2} & 0 & \widehat{\sigma}_1^{\,2} & \cdots & \widehat{\sigma}_{k-1}^{\,2} \\ 1 & \widehat{\sigma}_{k-1}^{\,2} & \widehat{\sigma}_k^{\,2} & & \vdots & \widehat{\sigma}_1^{\,2} & \ddots & & \vdots \\ \vdots & \vdots & \vdots & & \vdots & & & 0 & \widehat{\sigma}_1^{\,2} \\ 1 & \widehat{\sigma}_1^{\,2} & \widehat{\sigma}_2^{\,2} & \cdots & \widehat{\sigma}_k^{\,2} & \widehat{\sigma}_{k-1}^{\,2} & \cdots & \widehat{\sigma}_1^{\,2} & 0 \end{pmatrix} \qquad \text{in } \mathcal{T}^n$$

where $k = \left[\!\!\left[\dfrac{n}{2}\right]\!\!\right]$. If $n$ is odd, the entry $\widehat{\sigma}_k^{\,2}$ will be repeated twice, since $\widehat{\sigma}_k^{\,2} = \widehat{\sigma}_{k+1}^{\,2}$

The square of the volume of our chord $P$ is then expressed as:

$$V^2(P) = \frac{(-1)^n}{(n-1)! \cdot 2^{n-1}} \det\left(\widehat{B}(P)\right) \qquad \text{in } \mathcal{T}^n \qquad \text{(F.2)}$$

Comparing (F.1) and (F.2) we obtain an interesting relation between the determinant of the Hankel matrix and the Cayley-Menger determinant:

$$\left(\det\left(M_H(P)\right)\right)^2 = (-1)^n \cdot h^2 \cdot \det\left(\widehat{B}(P)\right)$$

with $h$ as above.

# Appendix C. Enrichment and Reduction between $\mathscr{T}^n$ and $\mathscr{T}^{n+1}$

Enrichment and reduction between $\mathscr{T}^2$ and $\mathscr{T}^3$

| | 3-01 | 3-02 | 3-02 I | 3-03 | 3-03 I | 3-06 | 3-04 | 3-04 I | 3-05 | 3-05 I | 3-07 | 3-07 I | 3-08 | 3-08 I | 3-09 | 3-10 | 3-11 | 3-11 I | 3-12 |
|---|---|---|---|---|---|---|---|---|---|---|---|---|---|---|---|---|---|---|---|
| 2-01 | △ | △ | △ | △ | △ |   | △ | △ | △ | △ |   |   |   |   |   |   |   |   |   |
| 2-02 | △ | △ | △ |   |   | △ |   |   |   |   | △ | △ | △ | △ | △ |   |   |   |   |
| 2-03 |   | △ | △ | △ | △ |   |   |   |   |   | △ | △ |   |   |   | △ | △ | △ |   |
| 2-04 |   |   |   | △ | △ | △ | △ | △ |   |   |   |   | △ | △ |   |   | △ | △ | △ |
| 2-05 |   |   |   |   |   |   | △ | △ | △ | △ | △ | △ |   |   | △ |   | △ | △ |   |
| 2-06 |   |   |   |   |   |   |   |   | △ | △ |   |   | △ | △ |   | △ |   |   |   |

Enrichment and reduction between $\mathcal{T}^3$ and $\mathcal{T}^4$

| | 3-01 | 3-02 | 3-02 I | 3-03 | 3-03 I | 3-06 | 3-04 | 3-04 I | 3-05 | 3-05 I | 3-07 | 3-07 I | 3-08 | 3-08 I | 3-09 | 3-10 | 3-11 | 3-11 I | 3-12 |
|---|---|---|---|---|---|---|---|---|---|---|---|---|---|---|---|---|---|---|---|
| 4-01 | ◁ | ◁ | ◁ | | | | | | | | | | | | | | | | |
| 4-02 | ◁ | ◁ | | | | ◁ | | | | | | | | | | | | | |
| 4-02 I | ◁ | | ◁ | ◁ | | ◁ | | | | | | | | | | | | | |
| 4-03 | | | ◁ | ◁ | ◁ | | | | | | | | | | | | | | |
| 4-04 | ◁ | | | ◁ | | | ◁ | | | | | | | | | | | | |
| 4-04 I | ◁ | | | | ◁ | | | ◁ | | | ◁ | ◁ | | | | | | | |
| 4-07 | | ◁ | | | | | ◁ | ◁ | | | | ◁ | | | | | | | |
| 4-11 | | ◁ | ◁ | | | ◁ | ◁ | | | | ◁ | | | | | | | | |
| 4-11 I | | | ◁ | | ◁ | ◁ | | ◁ | | | | ◁ | | | | | | | |
| 4-10 | | | | | | | | | | | ◁ | ◁ | | | | | | | |
| 4-05 | ◁ | | | | | | | | ◁ | | | | | | | | | | |
| 4-05 I | ◁ | | | | | | | | | ◁ | | | ◁ | | | | | | |
| 4-08 | | | | | | | | | ◁ | ◁ | | | | ◁ | | | | | |
| 4-06 | ◁ | | | | | | | | | ◁ | | | | | ◁ | | | | |
| 4-09 | | | | | | | | | ◁ | | | | | | | | | | |
| 4-13 | | ◁ | | ◁ | | | ◁ | ◁ | ◁ | ◁ | ◁ | ◁ | ◁ | ◁ | | ◁ | | | |
| 4-13 I | | | ◁ | | ◁ | | | | | ◁ | | ◁ | ◁ | | | ◁ | | | |
| 4-15Z | | | | | | | | | | | | | | | | | | | |
| 4-15Z I | | | | | | | | | ◁ | ◁ | | | ◁ | ◁ | | | | | |
| 4-12 | | | | ◁ | | | | | | ◁ | ◁ | | | | | | | | |
| 4-12 I | | ◁ | | | ◁ | | | | ◁ | ◁ | | | | | | | | | |
| 4-21 | | | | | | ◁ | | | ◁ | ◁ | | | | | | | | | |
| 4-29Z | | | | | | | | | ◁ | | | | | | | | ◁ | | |
| 4-29Z I | | | ◁ | | | | | | | ◁ | | | | | | | | ◁ | |
| 4-16 | | ◁ | | | | | ◁ | ◁ | | | | | | ◁ | | ◁ | ◁ | | |
| 4-16 I | | | | | | | | | | | | | | | ◁ | | ◁ | ◁ | |
| 4-14 | | | | | | | ◁ | ◁ | | | | | | ◁ | ◁ | ◁ | | ◁ | |
| 4-14 I | | | | | | | | | | | | | | | | | ◁ | ◁ | |
| 4-18 | | | | ◁ | ◁ | | | | ◁ | ◁ | | | ◁ | ◁ | | | ◁ | ◁ | |
| 4-18 I | | | | ◁ | ◁ | | | ◁ | | | | ◁ | ◁ | ◁ | | ◁ | ◁ | ◁ | |
| 4-17 | | | | ◁ | ◁ | | | ◁ | | | | | | | | | ◁ | | |
| 4-19 | | | | | | | | | | | ◁ | ◁ | | | ◁ | | ◁ | ◁ | ◁ |
| 4-19 I | | | | | | | | | | | ◁ | ◁ | | | ◁ | | ◁ | ◁ | ◁ |
| 4-22 | | | | | | ◁ | | | | | ◁ | ◁ | | | | ◁ | | | |
| 4-22 I | | | | | | ◁ | | | | | ◁ | ◁ | | | | ◁ | ◁ | | |
| 4-23 | | | | | ◁ | | | | | | | | | | | ◁ | | | |
| 4-20 | | | | | | ◁ | ◁ | | | | | | | | | | ◁ | ◁ | |
| 4-24 | | | | | | | | | | | | | | | | | | | ◁ |
| 4-25 | | | | | | | | | | | | | | | | | ◁ | | |
| 4-27 | | | | | | | | | | | | | | | | | | ◁ | |
| 4-27 I | | | | | | | | | | | | | | | | | | | |
| 4-26 | | | | | | | | | | | | | | | | | ◁ | | |
| 4-28 | | | | | | | | | | | | | | | | | | ◁ | |

91

# Enrichment and reduction between $\mathscr{T}^4$ and $\mathscr{T}^5$

| | 4-1 | 4-2 | 4-2 I | 4-3 | 4-4 | 4-4 I | 4-7 | 4-11 | 4-11 I | 4-1 | 4-5 | 4-5 I | 4-8 | 4-6 | 4-9 | 4-13 | 4-13 I | 4-15Z | 4-15Z I | 4-12 | 4-12 I |
|---|---|---|---|---|---|---|---|---|---|---|---|---|---|---|---|---|---|---|---|---|---|
| 5-1 | △ | △ | △ | △ | | | | | | | | | | | | | | | | | |
| 5-2 | △ | △ | | | △ | | | △ | | △ | | | | | | | | | | | |
| 5-2 I | △ | | | △ | | △ | | | △ | △ | | | | | | | | | | | |
| 5-3 | | △ | | △ | △ | | △ | △ | | | | | | | | | | | | | |
| 5-3 I | | | | △ | △ | △ | △ | △ | | | | | | | | | | | | | |
| 5-4 | △ | | | | △ | | | | | | △ | | | | | △ | | | | △ | |
| 5-4 I | △ | | | | | △ | | | | | | △ | | | | | △ | | | | △ |
| 5-6 | | | | | △ | | △ | | | | △ | | △ | | | | | △ | | | |
| 5-6 I | | | | | | △ | △ | | | | | △ | △ | | | | | △ | | | |
| 5-9 | | △ | | | | | | △ | | | △ | | | | | | | △ | | | |
| 5-9 I | | | | △ | | | | | △ | | | △ | | | | | | | | △ | |
| 5-1 | | | | △ | | | | | | △ | | | | | | △ | | △ | | | △ |
| 5-1 I | | | | △ | | | | | | △ | | | | | | | △ | △ | | △ | |
| 5-8 | | △ | △ | | | | | | | | | | | | | | | | | △ | △ |
| 5-12Z | | | | | | | | △ | △ | | | | △ | | | △ | △ | | | | |
| 5-5 | △ | | | | | | | | | | △ | | △ | | | | | | | | |
| 5-5 I | △ | | | | | | | | | | | △ | △ | | | | | | | | |
| 5-7 | | | | | | | | | | | △ | | △ | △ | △ | | | | | | |
| 5-7 I | | | | | | | | | | | | △ | △ | △ | △ | | | | | | |
| 5-14 | | | | | △ | | | | | | | | △ | | | | | △ | | | |
| 5-14 I | | | | | | △ | | | | | | | △ | | | | | | △ | | |
| 5-16 | | | | △ | | | | | | | | | | | | | | | | △ | |
| 5-16 I | | | | △ | | | | | | | | | | | | | | | | | △ |
| 5-18Z | | | | | | | △ | | | | | | | | | | | | | | △ |
| 5-18Z I | | | | | | | △ | | | | | | | | | | | | △ | | |
| 5-36Z | | △ | | | | | | | | | | | | △ | | △ | | | | | |
| 5-36Z I | | | △ | | | | | | | | | | | △ | | | △ | | | | |
| 5-11 | | | △ | | △ | | | | | | | | | | | | | | | | |
| 5-11 I | | △ | | | | △ | | | | | | | | | | | | | | | |
| 5-19 | | | | | | | | | | | | | | | △ | △ | | △ | | | |
| 5-19 I | | | | | | | | | | | | | | | △ | | △ | | △ | | |
| 5-15 | | | | | | | | | | | △ | △ | | | | | | | | | |
| 5-17Z | | | | △ | | | | | | | | | | | | | | | | | |
| 5-23 | | | | | | | | △ | | △ | | | | | | | | | | | |
| 5-23 I | | | | | | | | | △ | △ | | | | | | | | | | | |
| 5-24 | | | | | | | | △ | | | | | | | | | | | | | |
| 5-24 I | | | | | | | | | △ | | | | | | | | | | | | |
| 5-13 | | △ | | | | | | | | | | | | △ | | | | | | | |
| 5-13 I | | | △ | | | | | | | △ | | | | | | | | | | | |
| 5-2 | | | | | | | | | | | | | △ | | | | | | | | |
| 5-2 I | | | | | | | | | | | | | △ | | | | | | | | |
| 5-21 | | | | | | | △ | | | | | | | | | | | | | | |
| 5-21 I | | | | | | | △ | | | | | | | | | | | | | | |
| 5-38Z | | | | | △ | | | | | | | | | △ | | | | | | | |
| 5-38Z I | | | | | | △ | | | | | △ | | | | | | | | | | |
| 5-22 | | | | | | | | | | | | | | | △ | | | | | | |
| 5-37Z | | | | | △ | △ | | | | | | | | | | | | | | | |
| 5-27 | | | | | | | | △ | | | | | | | | | | | | | |
| 5-27 I | | | | | | | | | △ | | | | | | | | | | | | |
| 5-28 | | | | | | | | | | | | | | | | | | △ | | △ | |
| 5-28 I | | | | | | | | | | | | | | | | | | | △ | | △ |
| 5-26 | | | | | | | | | △ | | | | | | | | | | | △ | |
| 5-26 I | | | | | | | | △ | | | | | | | | | | | | | △ |
| 5-29 | | | | | | | | | | | | | | | | △ | | | | | |
| 5-29 I | | | | | | | | | | | | | | | | | △ | | | | |
| 5-25 | | | | | | | | | | △ | | | | | | △ | | | | | |
| 5-25 I | | | | | | | | | | △ | | | | | | | △ | | | | |
| 5-3 | | | | | | | | | | | | | | | | | | △ | | | |
| 5-3 I | | | | | | | | | | | | | | | | | | | △ | | |
| 5-33 | | | | | | | | | | | | | | | | | | | | | |
| 5-31 | | | | | | | | | | | | | | | | △ | | | | | △ |
| 5-31 I | | | | | | | | | | | | | | | | | △ | | | △ | |
| 5-32 | | | | | | | | | | | | | | | | | | △ | | | |
| 5-32 I | | | | | | | | | | | | | | | | | | | △ | | |
| 5-34 | | | | | | | | | | | | | | | | | | | | | |
| 5-35 | | | | | | | | | | | | | | | | | | | | | |

|  | 4-21 | 4-29Z | 4-29Z I | 4-16 | 4-16 I | 4-14 | 4-14 I | 4-18 | 4-18 I | 4-17 | 4-19 | 4-19 I | 4-22 | 4-22 I | 4-23 | 4-2 | 4-24 | 4-25 | 4-27 | 4-27 I | 4-26 | 4-28 |
|---|---|---|---|---|---|---|---|---|---|---|---|---|---|---|---|---|---|---|---|---|---|---|
| 5-1 |  |  |  |  |  |  |  |  |  |  |  |  |  |  |  |  |  |  |  |  |  |  |
| 5-2 |  |  |  |  |  |  |  |  |  |  |  |  |  |  |  |  |  |  |  |  |  |  |
| 5-2 I |  |  |  |  |  |  |  |  |  |  |  |  |  |  |  |  |  |  |  |  |  |  |
| 5-3 |  |  |  |  |  |  |  |  |  |  |  |  |  |  |  |  |  |  |  |  |  |  |
| 5-3 I |  |  |  |  |  |  |  |  |  |  |  |  |  |  |  |  |  |  |  |  |  |  |
| 5-4 |  |  |  |  |  |  |  |  |  |  |  |  |  |  |  |  |  |  |  |  |  |  |
| 5-4 I |  |  |  |  |  |  |  |  |  |  |  |  |  |  |  |  |  |  |  |  |  |  |
| 5-6 |  |  |  |  |  |  |  |  |  |  |  |  |  |  |  |  |  |  |  |  |  |  |
| 5-6 I |  |  |  |  |  |  |  |  |  |  |  |  |  |  |  |  |  |  |  |  |  |  |
| 5-9 | △ |  |  |  |  |  |  |  |  |  |  |  |  |  |  |  |  |  |  |  |  |  |
| 5-9 I | △ |  |  |  |  |  |  |  |  |  |  |  |  |  |  |  |  |  |  |  |  |  |
| 5-1 |  |  |  |  |  |  |  |  |  |  |  |  |  |  |  |  |  |  |  |  |  |  |
| 5-1 I |  |  |  |  |  |  |  |  |  |  |  |  |  |  |  |  |  |  |  |  |  |  |
| 5-8 | △ |  |  |  |  |  |  |  |  |  |  |  |  |  |  |  |  |  |  |  |  |  |
| 5-12Z |  |  |  |  |  |  |  |  |  |  |  |  |  |  |  |  |  |  |  |  |  |  |
| 5-5 |  | △ |  |  |  | △ |  |  |  |  |  |  |  |  |  |  |  |  |  |  |  |  |
| 5-5 I |  |  | △ |  |  |  | △ |  |  |  |  |  |  |  |  |  |  |  |  |  |  |  |
| 5-7 |  |  |  |  | △ |  |  |  |  |  |  |  |  |  |  |  |  |  |  |  |  |  |
| 5-7 I |  |  |  | △ |  |  |  |  |  |  |  |  |  |  |  |  |  |  |  |  |  |  |
| 5-14 |  |  |  | △ |  |  |  |  |  |  |  |  |  |  | △ |  |  |  |  |  |  |  |
| 5-14 I |  |  |  |  | △ |  |  |  |  |  |  |  |  |  | △ |  |  |  |  |  |  |  |
| 5-16 |  | △ |  |  |  |  |  | △ |  | △ |  |  |  |  |  |  |  |  |  |  |  |  |
| 5-16 I |  |  | △ |  |  |  |  |  | △ | △ |  |  |  |  |  |  |  |  |  |  |  |  |
| 5-18Z |  |  |  | △ |  |  | △ | △ |  |  |  |  |  |  |  |  |  |  |  |  |  |  |
| 5-18Z I |  |  |  |  | △ | △ |  |  | △ |  |  |  |  |  |  |  |  |  |  |  |  |  |
| 5-36Z |  |  |  |  |  |  |  | △ |  |  |  |  | △ |  |  |  |  |  |  |  |  |  |
| 5-36Z I |  |  |  |  |  |  |  |  | △ |  |  |  |  | △ |  |  |  |  |  |  |  |  |
| 5-11 |  |  |  |  |  | △ |  |  |  |  | △ |  | △ |  |  |  |  |  |  |  |  |  |
| 5-11 I |  |  |  |  |  |  | △ |  |  |  | △ |  |  | △ |  |  |  |  |  |  |  |  |
| 5-19 |  | △ |  |  |  |  |  |  |  |  | △ |  |  |  |  |  |  |  |  |  |  |  |
| 5-19 I |  |  | △ |  |  |  |  | △ |  |  |  |  |  |  |  |  |  |  |  |  |  |  |
| 5-15 |  |  |  | △ | △ |  |  |  |  |  |  |  |  |  |  |  |  | △ |  |  |  |  |
| 5-17Z |  |  |  |  |  | △ | △ |  |  |  | △ | △ |  |  |  |  |  |  |  |  |  |  |
| 5-23 |  |  |  |  |  | △ |  |  |  |  |  |  |  | △ | △ |  |  |  |  |  |  |  |
| 5-23 I |  |  |  |  |  |  | △ |  |  |  |  |  | △ |  | △ |  |  |  |  |  |  |  |
| 5-24 | △ | △ |  | △ |  |  |  |  |  |  |  |  |  | △ |  |  |  |  |  |  |  |  |
| 5-24 I | △ |  | △ |  | △ |  |  |  |  |  |  |  | △ |  |  |  |  |  |  |  |  |  |
| 5-13 |  | △ |  |  |  |  |  |  |  |  | △ |  |  |  |  |  | △ |  |  |  |  |  |
| 5-13 I |  |  | △ |  |  |  |  |  |  |  |  | △ |  |  |  |  | △ |  |  |  |  |  |
| 5-2 |  | △ |  |  | △ |  | △ |  |  |  |  |  |  |  |  | △ |  |  |  |  |  |  |
| 5-2 I |  |  | △ | △ |  | △ |  |  |  |  |  |  |  |  |  | △ |  |  |  |  |  |  |
| 5-21 |  |  |  |  |  |  |  |  |  | △ | △ |  |  |  |  | △ |  |  |  |  |  |  |
| 5-21 I |  |  |  |  |  |  |  |  |  | △ |  | △ |  |  |  | △ |  |  |  |  |  |  |
| 5-38Z |  |  |  |  |  |  |  | △ |  |  |  |  |  |  |  | △ |  |  | △ |  |  |  |
| 5-38Z I |  |  |  |  |  |  |  |  | △ |  |  |  |  |  |  | △ |  |  |  |  | △ |  |
| 5-22 |  |  |  |  |  |  |  | △ | △ |  | △ | △ |  |  |  |  |  |  |  |  |  |  |
| 5-37Z |  |  |  |  |  |  |  |  |  |  |  |  | △ | △ |  |  |  |  |  |  | △ |  |
| 5-27 |  |  |  |  |  |  | △ |  |  |  |  |  | △ |  |  |  | △ |  |  |  | △ |  |
| 5-27 I |  |  |  |  |  | △ |  |  |  |  |  |  |  | △ |  |  | △ |  |  |  | △ |  |
| 5-28 |  |  | △ |  |  |  |  |  |  |  |  |  |  |  |  |  |  | △ |  | △ |  |  |
| 5-28 I |  | △ |  |  |  |  |  |  |  |  |  |  |  |  |  |  |  | △ | △ |  |  |  |
| 5-26 |  |  |  |  |  |  |  |  |  |  | △ |  |  |  |  |  | △ |  | △ |  |  |  |
| 5-26 I |  |  |  |  |  |  |  |  |  |  |  | △ |  |  |  |  | △ |  |  |  | △ |  |
| 5-29 |  |  |  | △ |  |  | △ |  |  |  |  |  |  |  | △ |  |  |  |  |  | △ |  |
| 5-29 I |  |  | △ |  |  | △ |  |  |  |  |  |  |  |  | △ |  |  |  | △ |  |  |  |
| 5-25 |  | △ |  |  |  |  |  |  |  |  |  |  |  |  |  |  |  |  | △ |  | △ |  |
| 5-25 I | △ |  |  |  |  |  |  |  |  |  |  |  |  |  |  |  |  |  |  | △ | △ |  |
| 5-3 |  |  |  | △ |  |  |  |  |  |  | △ |  | △ |  |  |  | △ |  |  |  |  |  |
| 5-3 I |  |  | △ |  |  |  |  |  |  |  |  |  |  |  | △ | △ | △ |  |  |  |  |  |
| 5-33 | △ |  |  |  |  |  |  |  |  |  |  |  |  |  |  |  | △ | △ |  |  |  |  |
| 5-31 |  |  |  |  |  |  |  |  | △ |  |  |  |  |  |  |  |  |  | △ |  |  | △ |
| 5-31 I |  |  |  |  |  |  |  | △ |  |  |  |  |  |  |  |  |  |  | △ |  |  | △ |
| 5-32 |  |  |  |  |  |  |  | △ | △ |  |  |  |  |  |  |  |  |  | △ |  | △ |  |
| 5-32 I |  |  |  |  |  |  |  | △ | △ |  |  |  |  |  |  |  |  |  |  | △ | △ |  |
| 5-34 | △ |  |  |  |  |  |  |  |  |  |  |  | △ | △ |  |  |  |  | △ |  | △ |  |
| 5-35 |  |  |  |  |  |  |  |  |  |  |  |  | △ | △ | △ |  |  |  |  |  | △ |  |

# Enrichment and reduction between $\mathcal{T}^5$ and $\mathcal{T}^6$

| | 5-01 | 5-02 | 5-02 I | 5-03 | 5-03 I | 5-04 | 5-04 I | 5-06 | 5-06 I | 5-09 | 5-09 I | 5-10 | 5-10 I | 5-08 | 5-12Z | 5-05 | 5-05 I | 5-07 | 5-07 I | 5-14 | 5-14 I |
|---|---|---|---|---|---|---|---|---|---|---|---|---|---|---|---|---|---|---|---|---|---|
| 6-01 | △ | △ | △ | △ | △ | | | | | | | | | | | | | | | | |
| 6-03Z | | △ | | △ | | △ | | △ | | | | | △ | | △ | | | | | | |
| 6-03Z I | | | △ | | △ | | △ | | △ | | | △ | | | △ | | | | | | |
| 6-02 | △ | △ | | | | △ | | | | △ | | | | △ | | △ | | | | | |
| 6-02 I | △ | | △ | | | | △ | | | | △ | | △ | △ | | | | | | | |
| 6-04Z | | | | △ | △ | | | △ | △ | △ | △ | | | | | | | | | | |
| 6-05 | | | | | | △ | | △ | | | | | | | | △ | | △ | | | |
| 6-05 I | | | | | | | △ | | △ | | | | | | | | △ | | △ | | |
| 6-36Z | △ | | | | | △ | | | | | | | | | | △ | | | | | |
| 6-36Z I | △ | | | | | | △ | | | | | | | | | | △ | | | | |
| 6-06Z | | | | | | | | △ | △ | | | | | | | | | | | △ | △ |
| 6-10Z | | | | | △ | | | | | | | | | △ | | | | | | | |
| 6-10Z I | | | | △ | | | | | | | | | | △ | | | | | | | |
| 6-09 | | △ | | | | | | | | △ | | | | | | △ | | | | △ | |
| 6-09 I | | | △ | | | | | | | | △ | | | | | | △ | | | | △ |
| 6-12Z | | | | | | | | | | △ | | | | | △ | △ | | | | | |
| 6-12Z I | | | | | | | | | | | △ | | | | △ | | △ | | | | |
| 6-11Z | | | | △ | | | | | | | | △ | | | | | | | | | |
| 6-11Z I | | | | | △ | | | | | | | | △ | | | | | | | | △ |
| 6-13Z | | | | | | | | | | | | △ | △ | | | | | | | | |
| 6-38Z | | | | | | | | | | | | | | | | △ | | △ | △ | △ | |
| 6-08 | | △ | △ | | | | | | | | | | | | | | | | | | |
| 6-37Z | △ | | | | | | | | | | | | | | | △ | △ | | | | |
| 6-07 | | | | | | | | | | | | | | | | | | | | △ | △ |
| 6-16 | | | | | | | | | △ | | | | | | | | | | | | |
| 6-16 I | | | | | | | | △ | | | | | | | | | | | | | |
| 6-14 | | | | | △ | | | | | | | | | | | | | | | | |
| 6-14 I | | | | △ | | | | | | | | | | | | | | | | | |
| 6-17Z | | | | | | | | | | | | | | | | | | | △ | | |
| 6-17Z I | | | | | | | | | | | | | | | | | | △ | | | |
| 6-41Z | | | | | | △ | | | | | | | | | | | △ | | | △ | |
| 6-41Z I | | | | | | | △ | | | | | | | | | | | | | | △ |
| 6-40Z | | △ | | | | | | | | | | | | | | | △ | | | | |
| 6-40Z I | | | △ | | | | | | | | | | | | | △ | | | | | |
| 6-19Z | | | | | | | | | | | | | | | | | | | | | |
| 6-19Z I | | | | | | | | | | | | | | | | | | | | | |
| 6-18 | | | | | | | | | | | | | | | | | | △ | △ | | |
| 6-18 I | | | | | | | | | | | | | | | | | | △ | | | △ |
| 6-15 | | | | △ | | | | | | | | | | | | | | | | | |
| 6-15 I | | | | | △ | | | | | | | | | | | | | | | | |
| 6-39Z | | | △ | | | △ | | | | | | | | | | | | | | | |
| 6-39Z I | | △ | | | | | △ | | | | | | | | | | | | | | |
| 6-43Z | | | | | | | | △ | | | | | | | | | | | | | |
| 6-43Z I | | | | | | | | | △ | | | | | | | | | | | | |
| 6-22 | | | | | | | | | | △ | | | | | | | | | | | |
| 6-22 I | | | | | | | | | | | △ | | | | | | | | | | |
| 6-25Z | | | | | | | | | | | | | | | △ | | | | | | |
| 6-25Z I | | | | | | | | | | | | | | | △ | | | | | | |
| 6-44Z | | | | | | | | △ | | | | | | | | | | | | | |
| 6-44Z I | | | | | | | | | △ | | | | | | | | | | | | |
| 6-24Z | | | | | | | | | | | | △ | | | | | | | | | |
| 6-24Z I | | | | | | | | | | | | | △ | | | | | | | | |
| 6-21 | | | | | | | | | | △ | | | | △ | | | | | | | |
| 6-21 I | | | | | | | | | | | △ | | | △ | | | | | | | |
| 6-26Z | | | | | | | | | | | | | | | | | | | | | |
| 6-23Z | | | | | | | | | | | | △ | △ | | | | | | | | |
| 6-42Z | | | | | | △ | △ | | | | | | | | | | | | | | |
| 6-20 | | | | | | | | | | | | | | | | | | | | | |
| 6-27 | | | | | | | | | | | | △ | | | | | | | | | |
| 6-27 I | | | | | | | | | | | | | △ | | | | | | | | |
| 6-47Z | | | | | | | | | | | | | | | | | | | | | △ |
| 6-47Z I | | | | | | | | | | | | | | | | | | | | △ | |
| 6-29Z | | | | | | | | | | | | | | | | | | | | | |
| 6-50Z | | | | | | | | | | | | | | | | | | | | | |
| 6-31 | | | | | | | | | | | | | | | | | | | | | |
| 6-31 I | | | | | | | | | | | | | | | | | | | | | |
| 6-28Z | | | | | | | | | | | | | | | △ | | | | | | |
| 6-46Z | | | | | | | | | | △ | | | | | | | | | | | |
| 6-46Z I | | | | | | | | | | | △ | | | | | | | | | | |
| 6-48Z | | | | | | | | | | | | | | | | | | | | △ | △ |
| 6-49Z | | | | | | | | | | | | | | | | | | | | | |
| 6-30 | | | | | | | | | | | | | | | | | | | | | |
| 6-30 I | | | | | | | | | | | | | | | | | | | | | |
| 6-45Z | | | | | | | | | | | | | | △ | | | | | | | |
| 6-34 | | | | | | | | | | | | | | | | | | | | | |
| 6-34 I | | | | | | | | | | | | | | | | | | | | | |
| 6-33 | | | | | | | | | | | | | | | | | | | | | |
| 6-33 I | | | | | | | | | | | | | | | | | | | | | |
| 6-32 | | | | | | | | | | | | | | | | | | | | | |
| 6-35 | | | | | | | | | | | | | | | | | | | | | |

| | 5-16 | 5-16 I | 5-18Z | 5-18Z I | 5-36Z | 5-36Z I | 5-11 | 5-11 I | 5-19 | 5-19 I | 5-15 | 5-17Z | 5-23 | 5-23 I | 5-24 | 5-24 I | 5-13 | 5-13 I | 5-20 | 5-20 I | 5-21 | 5-21 I |
|---|---|---|---|---|---|---|---|---|---|---|---|---|---|---|---|---|---|---|---|---|---|---|
| 6-01 | | | | | | | | | | | | | | | | | | | | | | |
| 6-03Z | | | | | | | | | | | | | | | | | | | | | | |
| 6-03Z I | | | | | | | | | | | | | | | | | | | | | | |
| 6-02 | | | | | | | | | | | | | | | | | | | | | | |
| 6-02 I | | | | | | | | | | | | | | | | | | | | | | |
| 6-04Z | | | | | | | | | | | | | | | | | | | | | | |
| 6-05 | | | | △ | | | | | △ | | | | | | | | | | | | | |
| 6-05 I | | | △ | | | | | | | △ | | | | | | | | | | | | |
| 6-36Z | △ | | | | △ | | △ | | | | | | | | | | | | | | | |
| 6-36Z I | | △ | | | | △ | | △ | | | | | | | | | | | | | | |
| 6-06Z | | | | | | | | | | | | | | | | | | | | | | |
| 6-10Z | △ | | △ | | | | | △ | | | | | | | △ | | | | | | | |
| 6-10Z I | | △ | | △ | | | △ | | | | | | | | | △ | | | | | | |
| 6-09 | | | | | | | | | | | | | △ | | △ | | | | | | | |
| 6-09 I | | | | | | | | | | | | | | △ | | △ | | | | | | |
| 6-12Z | | | | | △ | | | | | △ | | | | | | △ | | | | | | |
| 6-12Z I | | | | | | △ | | | △ | | | | | | △ | | | | | | | |
| 6-11Z | | | △ | | △ | | | | | | | | | | | | △ | | | | | |
| 6-11Z I | | | | △ | | △ | | | | | | | △ | | | | | | | | | |
| 6-13Z | △ | △ | | | | | | | △ | △ | | | | | | | | | | | | |
| 6-38Z | | | | | | | | | | | | | | | | | | | △ | △ | | |
| 6-08 | | | | | | | △ | △ | | | | | △ | △ | | | | | | | | |
| 6-37Z | | | | | | | | | | | | △ | | | | | △ | △ | | | | |
| 6-07 | | | | | | | | | | | △ | | | | | | | | | | | |
| 6-16 | | | | | | | | △ | | | | | | | | | △ | | △ | | △ | |
| 6-16 I | | | | | | | △ | | | | | | | | | | | △ | | △ | | △ |
| 6-14 | | | | | | | △ | | | | | △ | | | | | | | | | △ | |
| 6-14 I | | | | | | | | △ | | | | △ | | | | | | | | | | △ |
| 6-17Z | | | | | △ | | | | △ | | | | | | | | △ | | | | | |
| 6-17Z I | | | | | | △ | | | | △ | | | | | | | | △ | | | | |
| 6-41Z | | | | | | | | | | | △ | | | | | | | | | | | |
| 6-41Z I | | | | | | | | | | | △ | | | | | | | | | | | |
| 6-40Z | | | | | △ | | | | | | | | | | | | | | | | | |
| 6-40Z I | | | | | | △ | | | | | | | | | | | | | | | | |
| 6-19Z | △ | | | △ | | | | | | | | △ | | | | | | | △ | | | △ |
| 6-19Z I | | △ | △ | | | | | | | | | △ | | | | | | | | △ | △ | |
| 6-18 | | | | | | | | | | △ | | | | | | | | | | △ | | |
| 6-18 I | | | | | | | | | △ | | | | | | | | | △ | | | | |
| 6-15 | △ | | | | | | | | | | | | | | | | △ | | | | △ | |
| 6-15 I | | △ | | | | | | | | | | | | | | | | △ | | | | △ |
| 6-39Z | | | | | | | | | | | | | | | | | | | △ | | | |
| 6-39Z I | | | | | | | | | | | | | | | | | △ | | | | | |
| 6-43Z | | | △ | | | | | | | | △ | | | | △ | | | | | | | |
| 6-43Z I | | | | △ | | | | | | | △ | | | | | | | | | △ | | |
| 6-22 | | | | | | | | | | | △ | | | | △ | | △ | | | | | |
| 6-22 I | | | | | | | | | | | △ | | | | | △ | | △ | | | | |
| 6-25Z | | | | | | | | | | | | | | △ | | | | | △ | | | |
| 6-25Z I | | | | | | | | | | | | | △ | | | | | | | △ | | |
| 6-44Z | | | | | | | | | | | | | | | | | | | | | △ | |
| 6-44Z I | | | | | | | | | | | | | | | | | | | | | | △ |
| 6-24Z | | | | | | | | | | | △ | | △ | | | | | | | | | |
| 6-24Z I | | | | | | | | | | | △ | | | △ | | | | | | | | |
| 6-21 | | | | | | | | | | | | | | | | | | △ | | | | |
| 6-21 I | | | | | | | | | | | | | | | | | △ | | | | | |
| 6-26Z | | | | | | | | | | | | | | | △ | | △ | | △ | △ | | |
| 6-23Z | | | | | | | | | | | | | | | | | | | | | | |
| 6-42Z | | | | | | | | | | | | | | | | | | | | | | |
| 6-20 | | | | | | | | | | | | | | | | | | | | | △ | △ |
| 6-27 | | △ | | | | | | | | | | | | | | | | | | | | |
| 6-27 I | △ | | | | | | | | | | | | | | | | | | | | | |
| 6-47Z | | | | | △ | | | △ | | | | | | | | | | | | | | |
| 6-47Z I | | | | | | △ | △ | | | | | | | | | | | | | | | |
| 6-29Z | | | △ | △ | | | | | | | | | | | | | | | | | | |
| 6-50Z | | | | | | | | | △ | △ | | | | | | | | | | | | |
| 6-31 | | | △ | | | | | | | | | | | | | | | | | | | △ |
| 6-31 I | | | | △ | | | | | | | | | | | | | | | | | △ | |
| 6-28Z | | | | | | | | | | | | | | | | | | | | | | |
| 6-46Z | | | | | | △ | | | | | | | | | | | | | | | | |
| 6-46Z I | | | | | | | △ | | | | | | | | | | | | | | | |
| 6-48Z | | | | | | | | | | | | | | | | | | | | | | |
| 6-49Z | △ | △ | | | | | | | | | | | | | | | | | | | | |
| 6-30 | | | | | | | | | △ | | | | | | | | | | | | | |
| 6-30 I | | | | | | | | | | △ | | | | | | | | | | | | |
| 6-45Z | | | | | △ | △ | | | | | | | | | | | | | | | | |
| 6-34 | | | | | | | | | | | | | | | △ | | | | | | | |
| 6-34 I | | | | | | | | | | | | | | | | △ | | | | | | |
| 6-33 | | | | | | | | | | | | | △ | | △ | | | | | | | |
| 6-33 I | | | | | | | | | | | | | | △ | | △ | | | | | | |
| 6-32 | | | | | | | | | | | | | △ | △ | | | | | | | | |
| 6-35 | | | | | | | | | | | | | | | | | | | | | | |

| | 5-38Z | 5-38Z | 5-22 | 5-37Z | 5-27 | 5-27 I | 5-28 | 5-28 I | 5-26 | 5-26 I | 5-29 | 5-29 I | 5-25 | 5-25 I | 5-30 | 5-30 I | 5-33 | 5-31 | 5-31 I | 5-32 | 5-32 I | 5-34 | 5-35 |
|---|---|---|---|---|---|---|---|---|---|---|---|---|---|---|---|---|---|---|---|---|---|---|---|
| 6-01 | | | | | | | | | | | | | | | | | | | | | | | |
| 6-03Z | | | | | | | | | | | | | | | | | | | | | | | |
| 6-03Z I | | | | | | | | | | | | | | | | | | | | | | | |
| 6-02 | | | | | | | | | | | | | | | | | | | | | | | |
| 6-02 I | | | | | | | | | | | | | | | | | | | | | | | |
| 6-04Z | | | | | | | | | | | | | | | | | | | | | | | |
| 6-05 | | | | | | | | | | | | | | | | | | | | | | | |
| 6-05 I | | | | | | | | | | | | | | | | | | | | | | | |
| 6-36Z | | | | | | | | | | | | | | | | | | | | | | | |
| 6-36Z I | | | | | | | | | | | | | | | | | | | | | | | |
| 6-06 | | | | | | | | | | | | | | | | | | | | | | | |
| 6-10Z | | | | | | | | | | | | | | | | | | | | | | | |
| 6-10Z I | | | | | | | | | | | | | | | | | | | | | | | |
| 6-09 | | | | | | | | | | | | | | | | | | | | | | | |
| 6-09 I | | | | | | | | | | | | | | | | | | | | | | | |
| 6-12Z | | | | | | | | | | | | | | | | | | | | | | | |
| 6-12Z I | | | | | | | | | | | | | | | | | | | | | | | |
| 6-11Z | | | | | | | | | | | | | | | | | | | | | | | |
| 6-11Z I | | | | | | | | | | | | | | | | | | | | | | | |
| 6-13Z | | | | | | | | | | | | | | | | | | | | | | | |
| 6-38Z | | | | | | | | | | | | | | | | | | | | | | | |
| 6-08 | | | | | | | | | | | | | | | | | | | | | | | |
| 6-37Z | | | | | | | | | | | | | | | | | | | | | | | |
| 6-07 | | | | | | | | | | | | | | | | | | | | | | | |
| 6-16 | | | | | | | | | | | | | | | △ | | | | | | | | |
| 6-16 I | | | | | | | | | | | | | | | | △ | | | | | | | |
| 6-14 | | | | △ | △ | | | | | | | | | | | | | | | | | | |
| 6-14 I | | | | △ | | △ | | | | | | | | | | | | | | | | | |
| 6-17Z | | | △ | | | | | | | | | | | | | △ | | | | | | | |
| 6-17Z I | | | △ | | | | | | | | | | | | △ | | | | | | | | |
| 6-41Z | | | | | | | △ | | | | △ | | | | | | | | | | | | |
| 6-41Z I | | | | | | | | △ | | | | △ | | | | | | | | | | | |
| 6-40Z | △ | | | | △ | | | | | | | | △ | | | | | | | | | | |
| 6-40Z I | | △ | | | | △ | | | | | | | | △ | | | | | | | | | |
| 6-19Z | | | △ | | | | | | | | | | | | | | | | | | | | |
| 6-19Z I | | | △ | | | | | | | | | | | | | | | | | | | | |
| 6-18 | △ | | | | | | | | | | △ | | | | | | | | | | | | |
| 6-18 I | | △ | | | | | | | | | △ | | | | | | | | | | | | |
| 6-15 | △ | | | | | | | | △ | | | | | | | | | | | | | | |
| 6-15 I | | △ | | | | | | | | △ | | | | | | | | | | | | | |
| 6-39Z | | | | △ | | | | | △ | | | | △ | | | | | | | | | | |
| 6-39Z I | | | | △ | | | | | | △ | | | | △ | | | | | | | | | |
| 6-43Z | △ | | | | | | | △ | | | | | | | | | | | | | | | |
| 6-43Z I | | △ | | | | | | △ | | | | | | | | | | | | | | | |
| 6-22 | | | | | | | | | | | | | | | △ | | △ | | | | | | |
| 6-22 I | | | | | | | | | | | | | | | | △ | △ | | | | | | |
| 6-25Z | | | | | △ | | | | | | △ | | △ | | | | | | | | | | |
| 6-25Z I | | | | | | △ | | | | | | △ | △ | | | | | | | | | | |
| 6-44Z | | △ | △ | △ | | | | | | | | | | | | | | | | | | △ | |
| 6-44Z I | △ | | △ | △ | | | | | | | | | | | | | | | | △ | | | |
| 6-24Z | | | | | | | △ | | △ | | | | | | △ | | | | | | | | |
| 6-24Z I | | | | | | | | △ | | | △ | | | | | △ | | | | | | | |
| 6-21 | | | | | | | △ | | △ | | | | | | | | | △ | | | | | |
| 6-21 I | | | | | | | | △ | △ | | | | | | | | | △ | | | | | |
| 6-26Z | | | | | △ | △ | | | | | | | | | | | | | | | | | |
| 6-23Z | | | | | | | △ | △ | | | | | △ | △ | | | | | | | | | |
| 6-42Z | △ | △ | | | | | | | | | | | | | | | | △ | △ | | | | |
| 6-20 | | | | | | | | | | | | | | | | | | | | | | | |
| 6-27 | | | | | | | | | | | | | △ | | | | | △ | | △ | | | |
| 6-27 I | | | | | | | | | | | | | △ | | | | | | △ | | △ | | |
| 6-47Z | | | | | | | | | △ | | | | | | | | | | | | △ | | △ |
| 6-47Z I | | | | | | | | | | △ | | | | | | | | | | △ | | | △ |
| 6-29Z | | | | | | | | | | | △ | △ | | | | | | △ | △ | | | | |
| 6-50Z | | | | | | | | | | | △ | △ | | | | | | | | △ | △ | | |
| 6-31 | | | | | △ | | | | △ | | | | | | | △ | | | | | △ | | |
| 6-31 I | | | | | | △ | | | △ | | | | | | | | | △ | | △ | | | |
| 6-28Z | | | △ | | | | | | △ | △ | | | | | | | | △ | △ | | | | |
| 6-46Z | | △ | | | △ | | | | | | | | | | | | | | | △ | | △ | |
| 6-46Z I | △ | | | | △ | | | | | | | | | | | | | | | | △ | △ | |
| 6-48Z | | | | △ | | | | | | | | | | | △ | △ | | | | | | | △ |
| 6-49Z | | | | | | | △ | △ | | | | | | | | | | | | △ | △ | | |
| 6-30 | | | | | | | | △ | | | | | | | | | | △ | | | | | |
| 6-30 I | | | | | | | △ | | | | | | | | | | | | △ | | | | |
| 6-45Z | | | | | | | | | | | | | | | | | | △ | △ | | | △ | |
| 6-34 | | | | | | | △ | | △ | | | | | | | △ | △ | | | | | △ | |
| 6-34 I | | | | | | | | △ | | △ | | | | | △ | | △ | | | | | △ | |
| 6-33 | | | | | | | | | | | | | △ | △ | | | | | | | | △ | △ |
| 6-33 I | | | | | | | | | △ | | | | △ | | | | | | | | | △ | △ |
| 6-32 | | | | | △ | △ | | | | | | | | | | | | | | | | | △ |
| 6-35 | | | | | | | | | | | | | | | | | | △ | | | | | |

# Enrichment and reduction between $\mathcal{T}^6$ and $\mathcal{T}^7$

| | 7-01 | 7-02 | 7-02 I | 7-04 | 7-04 I | 7-05 | 7-05 I | 7-03 | 7-03 I | 7-06 | 7-06 I | 7-07 | 7-07 I | 7-11 | 7-11 I | 7-36Z | 7-36Z | 7-14 | 7-14 I | 7-09 | 7-09 I | 7-01 |
|---|---|---|---|---|---|---|---|---|---|---|---|---|---|---|---|---|---|---|---|---|---|---|
| 6-01 | △ | △ | △ | | | | | △ | △ | | | | | | | | | | | | | △ |
| 6-03Z | △ | | | △ | | △ | | | | | | | | | △ | △ | | | | | | △ |
| 6-03Z I | △ | | | | △ | | △ | | | | | | | △ | | | △ | | | | | △ |
| 6-02 | △ | △ | | △ | | | | | | | | | | | | | | | | △ | | △ |
| 6-02 I | △ | | △ | | △ | | | | | | | | | | | | | | | | △ | △ |
| 6-04Z | △ | | | | | △ | △ | | | | | | | | | | | | | | | △ |
| 6-05 | | | | △ | | △ | | | | △ | | △ | | | | | | | | | | |
| 6-05 I | | | | | △ | | △ | | | | △ | | △ | | | | | | | | | |
| 6-36Z | | △ | | △ | | | | △ | | △ | | | | | | | | | | | | |
| 6-36Z I | | | △ | | △ | | | | △ | | △ | | | | | | | | | | | |
| 6-06Z | | | | | | △ | △ | | | | △ | | △ | | | | | | | | | |
| 6-10Z | | △ | | △ | | | | | | | | | | | | | | | | | | |
| 6-10Z I | | | △ | △ | | | | | | | | | | | | | | | | | | |
| 6-09 | | △ | | | | △ | | | | | | | | | | | | △ | | △ | | |
| 6-09 I | | | △ | | | | △ | | | | | | | | | | | | △ | | △ | |
| 6-12Z | | | | △ | | | △ | | | | | | | | | | △ | | | | | |
| 6-12Z I | | | | | △ | △ | | | | | | | | | | | | △ | | | | |
| 6-11Z | | △ | | | | | △ | | | | | | | | | △ | | | | | | |
| 6-11Z I | | | △ | | | △ | | | | | | | | | | | △ | | | | | |
| 6-13Z | | | | △ | △ | | | | | | | | | | | | | | | | | |
| 6-38Z | | | | | | | | | | △ | | △ | △ | | | | | △ | △ | | | |
| 6-08 | | △ | △ | | | | | | | | | | | △ | △ | | | | | | | |
| 6-37Z | | | | | | | | △ | △ | △ | △ | | | | | | | | | △ | △ | |
| 6-07 | | | | | | | | | | | | △ | △ | | | | | | | | | |
| 6-16 | | | | | | | | | | | △ | | | △ | | | | | | | | |
| 6-16 I | | | | | | | | | | △ | | | | | △ | | | | | | | |
| 6-14 | | | | | | | | △ | | | | | | △ | | | | | | | | |
| 6-14 I | | | | | | | | | △ | | | | | | △ | | | | | | | |
| 6-17Z | | | | | | | | △ | | | | | | | | | | | | | | |
| 6-17Z I | | | | | | | | | △ | | | | | | | | | | | | | |
| 6-41Z | | | | | | | | | | | | △ | | | | △ | | | | △ | | |
| 6-41Z I | | | | | | | | | | | | | △ | | | | △ | | | | △ | |
| 6-40Z | | | | | | | | △ | | | | | | | | △ | | △ | | | | |
| 6-40Z I | | | | | | | | | △ | | | | | | | | △ | | △ | | | |
| 6-19Z | | | | | | | | | | △ | | | | | | | | | | | | |
| 6-19Z I | | | | | | | | | | | △ | | | | | | | | | | | |
| 6-18 | | | | | | | | | | | | | △ | | | | | △ | | | | |
| 6-18 I | | | | | | | | △ | | | | | | | | | | | △ | | | |
| 6-15 | | | | | | | | △ | | | | | | | | | | | | | | |
| 6-15 I | | | | | | | | | △ | | | | | | | | | | | | | |
| 6-39Z | | | | | | | | △ | | | | | | △ | | | | | | | | |
| 6-39Z I | | | | | | | | | △ | | | | | | △ | | | | | | | |
| 6-43Z | | | | | | | | | | | | | △ | | | △ | | | | | | |
| 6-43Z I | | | | | | | | | | | | △ | | | | | △ | | | | | |
| 6-22 | | | | | | | | | | | | | | | | | | | | △ | | |
| 6-22 I | | | | | | | | | | | | | | | | | | | | | △ | |
| 6-25Z | | | | | | | | | | | | | | △ | | △ | | △ | | | | |
| 6-25Z I | | | | | | | | | | | | | | | △ | | △ | △ | | | | |
| 6-44Z | | | | | | | | | | | | | | | | | | | | | | |
| 6-44Z I | | | | | | | | | | | | | | | | | | | | | | |
| 6-24Z | | | | | | | | | | | | | | △ | | | | | | △ | | |
| 6-24Z I | | | | | | | | | | | | | | | △ | | | | | | △ | |
| 6-21 | | | | | | | | | | | | | | | | | | | | △ | | |
| 6-21 I | | | | | | | | | | | | | | | | | | | | | △ | |
| 6-26Z | | | | | | | | | | | | | | | | | | △ | △ | | | |
| 6-23Z | | | | | | | | | | | | | | | | △ | △ | | | | | |
| 6-42Z | | | | | | | | | | | | | | | | | | | | | | |
| 6-20 | | | | | | | | | | | | | | | | | | | | | | |
| 6-27 | | | | | | | | | | | | | | | | | | | | | | |
| 6-27 I | | | | | | | | | | | | | | | | | | | | | | |
| 6-47Z | | | | | | | | | | | | | | | | | | | | | | |
| 6-47Z I | | | | | | | | | | | | | | | | | | | | | | |
| 6-29Z | | | | | | | | | | | | | | | | | | | | | | |
| 6-50Z | | | | | | | | | | | | | | | | | | | | | | |
| 6-31 | | | | | | | | | | | | | | | | | | | | | | |
| 6-31 I | | | | | | | | | | | | | | | | | | | | | | |
| 6-28Z | | | | | | | | | | | | | | | | | | | | | | |
| 6-46Z | | | | | | | | | | | | | | | | | | | | | | |
| 6-46Z I | | | | | | | | | | | | | | | | | | | | | | |
| 6-48Z | | | | | | | | | | | | | | | | | | | | | | |
| 6-49Z | | | | | | | | | | | | | | | | | | | | | | |
| 6-30 | | | | | | | | | | | | | | | | | | | | | | |
| 6-30 I | | | | | | | | | | | | | | | | | | | | | | |
| 6-45Z | | | | | | | | | | | | | | | | | | | | | | |
| 6-34 | | | | | | | | | | | | | | | | | | | | | | |
| 6-34 I | | | | | | | | | | | | | | | | | | | | | | |
| 6-33 | | | | | | | | | | | | | | | | | | | | | | |
| 6-33 I | | | | | | | | | | | | | | | | | | | | | | |
| 6-32 | | | | | | | | | | | | | | | | | | | | | | |
| 6-35 | | | | | | | | | | | | | | | | | | | | | | |

| | 7-13 | 7-13 I | 7-38Z | 7-38Z | 7-08 | 7-15 | 7-37Z | 7-16 | 7-16 I | 7-18Z | 7-18Z | 7-19 | 7-19 I | 7-10 | 7-10 I | 7-20 | 7-20 I | 7-21 | 7-21 I | 7-17Z | 7-12Z | 7-22 |
|---|---|---|---|---|---|---|---|---|---|---|---|---|---|---|---|---|---|---|---|---|---|---|
| 6-01 | | | | | | | | | | | | | | | | | | | | | | |
| 6-03Z | | | | | | | | △ | | | | | | | | | | | | | | |
| 6-03Z I | | | | | | | | | △ | | | | | | | | | | | | | |
| 6-02 | | | | | △ | | | | | | | | | △ | | | | | | | | |
| 6-02 I | | | | | △ | | | | | | | | | | △ | | | | | | | |
| 6-04Z | △ | △ | | | | | | | | | | | | | | | | | | △ | | |
| 6-05 | | | | | | | | | | | △ | △ | | | | | | | | | | |
| 6-05 I | | | | | | | | | | △ | | | △ | | | | | | | | | |
| 6-36Z | | | | | | | | | | | | | | | △ | | | | | | △ | |
| 6-36Z I | | | | | | | | | | | | | | △ | | | | | | | △ | |
| 6-06Z | | | | | | | | | | | | | | | | △ | △ | | | | | |
| 6-10Z | △ | | | | | | △ | | | | | | | | | | | | | | | |
| 6-10Z I | | △ | | | | | △ | | | | | | | | | | | | | | | |
| 6-09 | | | | | | | | | | | | | | | | | | | | | | |
| 6-09 I | | | | | | | | | | | | | | | | | | | | | | |
| 6-12Z | | | | | | △ | | | | | | | | | | | | | | | | |
| 6-12Z I | | | | | | △ | | | | | | | | | | | | | | | | |
| 6-11Z | | | △ | | | | | | | | | | | | | | | | | | | |
| 6-11Z I | | | | △ | | | | | | | | | | | | | | | | | | |
| 6-13Z | | | △ | △ | | | | | | | | | | | | | | | | | | |
| 6-38Z | | | | | | | | | | | | | | | | | | | | | | |
| 6-08 | | | | | | | | | | | | | | | | | | | | | | |
| 6-37Z | | | | | | | | | | | | | | | | | | | | | | |
| 6-07 | | | | | | △ | | | | | | | | | | | | | | | | |
| 6-16 | △ | | | | | | | | | | | | | | | △ | | △ | | | | |
| 6-16 I | | △ | | | | | | | | | | | | | | | △ | | △ | | | |
| 6-14 | | | | | | | △ | | | | | | | | | | | | △ | △ | | |
| 6-14 I | | | | | | | △ | | | | | | | | | | | △ | | △ | | |
| 6-17Z | | | △ | | | △ | | | | △ | | | | | | △ | | | | | | |
| 6-17Z I | | | | △ | | △ | | | | | △ | | | | | | △ | | | | | |
| 6-41Z | | | | | | | | | | | | △ | | | | | | | | | △ | |
| 6-41Z I | | | | | | | | | | | △ | | | | | | | | | | △ | |
| 6-40Z | | | | | | | | | | △ | | | | △ | | | | | | | | |
| 6-40Z I | | | | | | | | | | | △ | | | | △ | | | | | | | |
| 6-19Z | | | | △ | | | △ | | | | | | | | | | | △ | | | | △ |
| 6-19Z I | | | △ | | | | △ | | | | | | | | | | | | △ | | | △ |
| 6-18 | | | △ | | | | | | | | | | △ | | | | △ | | | | | |
| 6-18 I | | | | △ | | | | | | | | △ | | | | △ | | | | | | |
| 6-15 | △ | | △ | | | | | △ | | | | | | | | | | △ | | | | |
| 6-15 I | | △ | | △ | | | | | △ | | | | | | | | | | △ | | | |
| 6-39Z | | | | | △ | | | | △ | | △ | | | | | | | | | | | |
| 6-39Z I | | | | | △ | | | △ | | △ | | | | | | | | | | | | |
| 6-43Z | △ | | | | | | | | | | | △ | | | | | | | | | | △ |
| 6-43Z I | | △ | | | | | | | | | | | △ | | | | | | | | | △ |
| 6-22 | △ | | | | | △ | | | | | | | | | | | | | | | | |
| 6-22 I | | △ | | | | △ | | | | | | | | | | | | | | | | |
| 6-25Z | | | | | | | | | | | | | | | | | | | | | | |
| 6-25Z I | | | | | | | | | | | | | | | | | | | | | | |
| 6-44Z | | | | | | | | △ | | | △ | | | | | △ | | △ | | △ | | △ |
| 6-44Z I | | | | | | | | | △ | △ | | | | | | | △ | | △ | △ | | △ |
| 6-24Z | | | | △ | | | | | | | | | | | | | | | | | | |
| 6-24Z I | | | △ | | | | | | | | | | | | | | | | | | | |
| 6-21 | | △ | | | △ | | | | | | | | | | | | | | | | | |
| 6-21 I | △ | | | | △ | | | | | | | | | | | | | | | | | |
| 6-26Z | | | | | | | △ | | | | | | | | | | | | | | | |
| 6-23Z | | | | | △ | | | | | | | | | | | | | | | | | |
| 6-42Z | | | | | | | | △ | △ | | | △ | △ | △ | △ | | | | | | | |
| 6-20 | | | | | | | | | | | | | | | | | | △ | △ | | | |
| 6-27 | | | | | | | | | △ | | | | | △ | | | | | | | | |
| 6-27 I | | | | | | | | △ | | | | | | | △ | | | | | | | |
| 6-47Z | | | | | | | | | | | | | | | | | | △ | | | △ | |
| 6-47Z I | | | | | | | | | | | | | | | | | △ | | | | △ | |
| 6-29Z | | | | | | | | | | | | △ | △ | | | | | | | | | |
| 6-50Z | | | | | | | | | | △ | △ | | | | | | | | | | | |
| 6-31 | | | | | | | | | | △ | | | | | | | | | △ | | | |
| 6-31 I | | | | | | | | | | | △ | | | | | | | △ | | | | |
| 6-28Z | | | | | | | | △ | △ | | | | | | | | | | | | | |
| 6-46Z | | | | | | | | | | | | | | △ | | | | | | △ | | |
| 6-46Z I | | | | | | | | | | | | | | | △ | | | | | △ | | |
| 6-48Z | | | | | | | | | | | | | | | | △ | △ | | | | | |
| 6-49Z | | | | | | | | | | | | | | | | | | | | | △ | △ |
| 6-30 | | | | | | | | | | | | △ | | | | | | | | | | |
| 6-30 I | | | | | | | | | | | | | △ | | | | | | | | | |
| 6-45Z | | | | | | | | | | | | | | △ | △ | | | | | | | |
| 6-34 | | | | | | | | | | | | | | | | | | | | | | |
| 6-34 I | | | | | | | | | | | | | | | | | | | | | | |
| 6-33 | | | | | | | | | | | | | | | | | | | | | | |
| 6-33 I | | | | | | | | | | | | | | | | | | | | | | |
| 6-32 | | | | | | | | | | | | | | | | | | | | | | |
| 6-35 | | | | | | | | | | | | | | | | | | | | | | |

| | 7-27 | 7-27 I | 7-24 | 7-24 I | 7-28 | 7-28 I | 7-31 | 7-31 I | 7-25 | 7-25 I | 7-26 | 7-26 I | 7-30 | 7-30 I | 7-23 | 7-23 I | 7-29 | 7-29 I | 7-32 | 7-32 I | 7-33 | 7-34 | 7-35 |
|---|---|---|---|---|---|---|---|---|---|---|---|---|---|---|---|---|---|---|---|---|---|---|---|
| 6-01 | | | | | | | | | | | | | | | | | | | | | | | |
| 6-03Z | | | | | | | | | | | | | | | | | | | | | | | |
| 6-03Z | | | | | | | | | | | | | | | | | | | | | | | |
| 6-02 | | | | | | | | | | | | | | | | | | | | | | | |
| 6-02 I | | | | | | | | | | | | | | | | | | | | | | | |
| 6-04Z | | | | | | | | | | | | | | | | | | | | | | | |
| 6-05 | | | | | | | | | | | | | | | | | | | | | | | |
| 6-05 I | | | | | | | | | | | | | | | | | | | | | | | |
| 6-36Z | | | | | | | | | | | | | | | | | | | | | | | |
| 6-36Z | | | | | | | | | | | | | | | | | | | | | | | |
| 6-06Z | | | | | | | | | | | | | | | | | | | | | | | |
| 6-10Z | | | | | | | | | | △ | | | | | | | | | | | | | |
| 6-10Z | | | | | | | | | △ | | | △ | | | | | | | | | | | |
| 6-09 | | | △ | | | | | | | | | | | | | | | | | | | | |
| 6-09 I | | | | △ | | | | | | | | | | | △ | | | | | | | | |
| 6-12Z | | | | | | △ | | | | | | | | | | △ | | | | | | | |
| 6-12Z | | | | | △ | | | | | | | | | | | | | △ | | | | | |
| 6-11Z | △ | | | | | | | | △ | | | | | | | | | | | | | | |
| 6-11Z | | △ | | | | | | | | △ | | | | | | | | | | | | | |
| 6-13Z | | | | | | | △ | △ | | | | | | | | | | | | | | | |
| 6-38Z | | | | | | | | | | | | | | | | | | | | | | | |
| 6-08 | | | | | | | | | | | | | | | △ | △ | | | | | | | |
| 6-37Z | | | | | | | | | | | | | | | | | | | | | | | |
| 6-07 | | | | | | | | | | | | | | | | | | | | | | | |
| 6-16 | | | | | | | | | | | | | △ | | | | | | | | | | |
| 6-16 I | | | | | | | | | | | | | | △ | | | | | | | | | |
| 6-14 | | △ | | | | | | | | | | | | | | | | | | | | | |
| 6-14 I | △ | | | | | | | | | | | | | | | | | | | | | | |
| 6-17Z | | | | | △ | | | | | | | | | | | | | | | | | | |
| 6-17Z | | | | | | △ | | | | | | | | | | | | | | | | | |
| 6-41Z | | | | △ | | | | | | | | | | | | | | | | | | | |
| 6-41Z | | | △ | | | | | | | | | | | | | | | | | | | | |
| 6-40Z | | | | | | | | | | | | | | | | △ | | | | | | | |
| 6-40Z | | | | | | | | | | | | | | | △ | | | | | | | | |
| 6-19Z | | | | | | | | | | | | | | | | | | | | | △ | | |
| 6-19Z | | | | | | | | | | | | | | | | | | | | △ | | | |
| 6-18 | | | | | | | | | | | | | | | | | △ | | | | | | |
| 6-18 I | | | | | | | | | | | | | | | | △ | | | | | | | |
| 6-15 | | | | | | | | | | | △ | | | | | | | | | | | | |
| 6-15 I | | | | | | | | | | | △ | | | | | | | | | | | | |
| 6-39Z | | | | △ | | | | | | | | | | | | | | | | | | | |
| 6-39Z | | | △ | | | | | | | | | | | | | | | | | | | | |
| 6-43Z | | | | | | | | | | | | | | △ | | | | | | | | | |
| 6-43Z | | | | | | | | | | | | | △ | | | | | | | | | | |
| 6-22 | | | △ | | | | | | | | | | | | | | | | | | △ | | |
| 6-22 I | | | | △ | | | | | | | | | | △ | | | | | | | △ | | |
| 6-25Z | | | | | | | | | | | | | | | | | △ | | △ | | | △ | |
| 6-25Z | | | | | | | | | | | | | | | | | | △ | | △ | | △ | |
| 6-44Z | | | | | | | | | | | | | | | | | | | | | | | |
| 6-44Z | | | | | | | | | | | | | | | | | | | | | | | |
| 6-24Z | △ | | | | | | | | | | | | | | | | | | △ | | | △ | |
| 6-24Z | | △ | | | | | | | | | | | | | | | | | | △ | | △ | |
| 6-21 | | | | | | △ | | | | | △ | | | | | | | | | | | △ | |
| 6-21 I | | | | | △ | | | | | | | △ | | | | | | | | | | △ | |
| 6-26Z | | | | | | | | | | | | | △ | △ | | | | | | | | | △ |
| 6-23Z | | | | | | | △ | △ | | | | | | | | | | | | | | △ | |
| 6-42Z | | | | | | | | | | | | | | | | | | | | | | | |
| 6-20 | | | | | | | | | | | | | | | | | | | | | | | |
| 6-27 | | | | | | | △ | | △ | | | | | | | | | | △ | | | | |
| 6-27 I | | | | | | | | △ | | △ | | | | | | | | | △ | | | | |
| 6-47Z | △ | | | | | | | | | △ | | | | | △ | △ | | | | | | | |
| 6-47Z | | △ | | | | | | △ | | | | | | | △ | | | △ | | | | | |
| 6-29Z | | | | | | | | | △ | △ | | | | | | | | | △ | △ | | | |
| 6-50Z | | | | | | | △ | △ | | | | | | | | | △ | △ | | | | | |
| 6-31 | △ | | | | | | | | | | △ | | △ | | | | | | △ | | | | |
| 6-31 I | | △ | | | | | | | | | | △ | | | | | | | △ | | | | |
| 6-28Z | | | | | △ | △ | | | | | | | | | | | | | △ | △ | | | |
| 6-46Z | | | | | | | | | | | △ | | △ | | △ | | △ | | | | | | |
| 6-46Z | | | | | | | | | | | | △ | | △ | | △ | | △ | | | | | |
| 6-48Z | △ | △ | △ | △ | | | | | | | | | | | | | | | | | | | |
| 6-49Z | | | | | | | △ | △ | | | △ | △ | | | | | | | | | | | |
| 6-30 | | | | | △ | | △ | | | | | | | | | | | | | | | | |
| 6-30 I | | | | | | △ | | △ | | | | | | | | | | | | | | | |
| 6-45Z | | | | | △ | △ | | | △ | △ | | | | | | | | | | | | | |
| 6-34 | | | △ | | △ | | | | | | △ | | △ | | | | | | | | △ | △ | |
| 6-34 I | | | | △ | | △ | | | | | | △ | | | △ | | | | | | △ | △ | |
| 6-33 | | | △ | | | | | | △ | | | | | | △ | | △ | | | | | △ | △ |
| 6-33 I | | | | △ | | | | △ | | | | | | | △ | △ | | | | | | △ | △ |
| 6-32 | △ | △ | | | | | | | | | | | | | | | | | △ | △ | | | △ |
| 6-35 | | | | | | | | | | | | | | | | | | | | | | △ | |

Enrichment and reduction between $\mathcal{T}^7$ and $\mathcal{T}^8$

| | 8-01 | 8-02 | 8-02 I | 8-05 | 8-05 I | 8-04 | 8-04 I | 8-06 | 8-03 | 8-08 | 8-07 | 8-09 | 8-12 | 8-12 I | 8-15Z | 8-15Z I | 8-11 | 8-11 I | 8-19 | 8-19 I | 8-18 |
|---|---|---|---|---|---|---|---|---|---|---|---|---|---|---|---|---|---|---|---|---|---|
| 7-01 | △ | △ | △ | | | | | | △ | | | | | | | | | | | | |
| 7-02 | △ | △ | | | | △ | | | | | | | | | | | | △ | | | |
| 7-02 I | △ | | △ | | | | △ | | | | | | | | | | | | | | |
| 7-04 | △ | | | △ | | △ | | | | | | | | △ | | | | | | | |
| 7-04 I | △ | | | | △ | | △ | | | | | | △ | | | | | | | | |
| 7-05 | △ | | | △ | | | | △ | | | | | | | | | | | | | |
| 7-05 I | △ | | | | △ | | | △ | | | | | | | | | | | | | |
| 7-03 | | △ | | | | △ | | | △ | | △ | | | | | | | | | △ | |
| 7-03 I | | | △ | | | | △ | | △ | | △ | | | | | | △ | | | | |
| 7-06 | | | | △ | | △ | | | | △ | △ | | | | | △ | | | | | |
| 7-06 I | | | | | △ | | △ | | | △ | △ | | | | △ | | | | | | |
| 7-07 | | | | △ | | | | △ | △ | | | △ | | | | | | | | | |
| 7-07 I | | | | | △ | | | △ | △ | | | △ | | | | | | | | | |
| 7-11 | | △ | | | | △ | | | | | | | | | | | | | | | |
| 7-11 I | | | △ | △ | | | | | | | | | | | | | | | | | |
| 7-36Z | | △ | | | | | | △ | | | | | | | | | | | | | △ |
| 7-36Z I | | | △ | | | | | △ | | | | | | | | | | | | | |
| 7-14 | | | | | | △ | | △ | | | | | | | △ | | | | | | |
| 7-14 I | | | | | | | △ | △ | | | | | | | | △ | | | | | |
| 7-09 | | △ | | △ | | | | | | | | | | | △ | | △ | | | | |
| 7-09 I | | | △ | | △ | | | | | | | | | | | △ | | △ | | | |
| 7-13 | | △ | | | △ | | | | | | | | | | | | | | △ | | |
| 7-13 I | | | △ | △ | | | | | | | | | | | | | | | | △ | |
| 7-38Z | | | | | △ | △ | | | | | | | | | | | | | | | △ |
| 7-38Z I | | | △ | | | | △ | | | | | | | | | | | | | | |
| 7-08 | | △ | △ | | | | | | | | | | △ | △ | | | | | | | |
| 7-15 | | | | △ | △ | | | | | | | | | | | | | | | | |
| 7-37Z | | | | | | △ | △ | | | | | | | | | | | | △ | △ | |
| 7-16 | | | | | | | | | △ | | | | | △ | | | | | | | △ |
| 7-16 I | | | | | | | | | △ | | | | △ | | | | | | | | |
| 7-18Z | | | | | | | | | | | △ | | △ | | | | | | | | △ |
| 7-18Z I | | | | | | | | | | | △ | | | △ | | | | | | | |
| 7-19 | | | | | | | | | | | | △ | | | | △ | | | | | |
| 7-19 I | | | | | | | | | | | | △ | | | △ | | | | | | △ |
| 7-10 | | | | | | | | | △ | | | | △ | | △ | | | | | | |
| 7-10 I | | | | | | | | | △ | | | | | △ | | △ | | | | | |
| 7-20 | | | | | | | | | | △ | | | | | | | | | | | |
| 7-20 I | | | | | | | | | | △ | | | | | | | | | | | |
| 7-21 | | | | | | | | | | | △ | | | | | | | | △ | | |
| 7-21 I | | | | | | | | | | | △ | | | | | | | | | △ | |
| 7-17Z | | | | | | | | △ | | | | | | | | | | | △ | △ | |
| 7-12Z | | | | | | | | | | △ | | | | | | | △ | △ | | | |
| 7-22 | | | | | | | | | | △ | | | | | | | | | △ | △ | △ |
| 7-27 | | | | | | | | | | | | | | | | | △ | | | | |
| 7-27 I | | | | | | | | | | | | | | | | | | △ | | | |
| 7-24 | | | | | | | | | | | | | | | | | △ | | | | |
| 7-24 I | | | | | | | | | | | | | | | | | | △ | | | |
| 7-28 | | | | | | | | | | | | | △ | | | △ | | | | | |
| 7-28 I | | | | | | | | | | | | | | △ | △ | | | | | | |
| 7-31 | | | | | | | | | | | | | △ | | | | | | | | |
| 7-31 I | | | | | | | | | | | | | | △ | | | | | | | △ |
| 7-25 | | | | | | | | | | | | | | | | | | | | | |
| 7-25 I | | | | | | | | | | | | | | | | | | | | | |
| 7-26 | | | | | | | | | | | | | △ | | | | △ | | | △ | |
| 7-26 I | | | | | | | | | | | | | | △ | | | △ | △ | | | |
| 7-30 | | | | | | | | | | | | | | | △ | | | | △ | | |
| 7-30 I | | | | | | | | | | | | | | | | △ | | | | △ | |
| 7-23 | | | | | | | | | | | | | | | | | △ | | | | |
| 7-23 I | | | | | | | | | | | | | | | | | | △ | | | |
| 7-29 | | | | | | | | | | | | | | | | | | | | | |
| 7-29 I | | | | | | | | | | | | | | | | | | | | | |
| 7-32 | | | | | | | | | | | | | | | △ | | | | | | |
| 7-32 I | | | | | | | | | | | | | | | | △ | | | | | △ |
| 7-33 | | | | | | | | | | | | | | | | | | | | | |
| 7-34 | | | | | | | | | | | | | | | | | | | | | |
| 7-35 | | | | | | | | | | | | | | | | | | | | | |

| | 8-18 I | 8-16 | 8-16 I | 8-29Z | 8-29Z I | 8-10 | 8-13 | 8-13 I | 8-14 | 8-14 I | 8-20 | 8-17 | 8-22 | 8-22 I | 8-27 | 8-27 I | 8-24 | 8-23 | 8-21 | 8-26 | 8-25 | 8-28 |
|---|---|---|---|---|---|---|---|---|---|---|---|---|---|---|---|---|---|---|---|---|---|---|
| 7-01 | | | | | | | | | | | | | | | | | | | | | | |
| 7-02 | | | | | | △ | | | | | | | | | | | | | | | | |
| 7-02 I | | | | | | △ | | | | | | | | | | | | | | | | |
| 7-04 | | | | | | | △ | | | | | | | | | | | | | | | |
| 7-04 I | | | | | | | | △ | | | | | | | | | | | | | | |
| 7-05 | | | | △ | | | | | | △ | | | | | | | | | | | | |
| 7-05 I | | | | | △ | | | | △ | | | | | | | | | | | | | |
| 7-03 | | | | | | | | | | | | | | | | | | | | | | |
| 7-03 I | | | | | | | | | | | | | | | | | | | | | | |
| 7-06 | | | | | | | | | | | | | | | | | | | | | | |
| 7-06 I | | | | | | | | | | | | | | | | | | | | | | |
| 7-07 | | | △ | | | | | | | | | | | | | | | | | | | |
| 7-07 I | | △ | | | | | | | | | | | | | | | | | | | | |
| 7-11 | | | | | | | | | △ | | △ | | △ | | | | | | | | | |
| 7-11 I | | | | | | | | | | △ | △ | | △ | | | | | | | | | |
| 7-36Z | | | | | | △ | | | | | | | △ | | | | | | | | | |
| 7-36Z I | △ | | | | | | | △ | | | | △ | | | | | | | | | | |
| 7-14 | | △ | | | | | | | | | | | | | | | | | | △ | | |
| 7-14 I | | | △ | | | | | | | | | | | | | | | | | △ | | |
| 7-09 | | | | | | | | | | | | | | | | | | | △ | | | |
| 7-09 I | | | | | | | | | | | | | | | | | | | △ | | | |
| 7-13 | | | | △ | | | | | | | | | | | | | △ | | | | | |
| 7-13 I | | | | | △ | | | | | | | | | | | | △ | | | | | |
| 7-38Z | | | | | | | | | | | △ | | | | △ | | | | | | | |
| 7-38Z I | △ | | | | | | | | | | △ | | | | | △ | | | | | | |
| 7-08 | | | | | | | | | | | | | | | | | | | △ | | | |
| 7-15 | | △ | △ | | | | | | | | | | | | | | | | | | △ | |
| 7-37Z | | | | | | | | | | | | | | | | | | | | △ | | |
| 7-16 | | | | △ | | | | | | | △ | | | | | | | | | | | |
| 7-16 I | △ | | | | △ | | | | | | △ | | | | | | | | | | | |
| 7-18Z | | △ | | | | | | | △ | | | | | | | | | | | | | |
| 7-18Z I | △ | | △ | | | | | | | △ | | | | | | | | | | | | |
| 7-19 | △ | | | △ | | | △ | | | | | | | | | | | | | | | |
| 7-19 I | | | | | △ | | | △ | | | | | | | | | | | | | | |
| 7-10 | | | | | | △ | △ | | | | | | | | | | | | | | | |
| 7-10 I | | | | | | △ | | △ | | | | | | | | | | | | | | |
| 7-20 | | | △ | △ | | | | | △ | | △ | | | | | | | | | | | |
| 7-20 I | | △ | | △ | | | | | | △ | △ | | | | | | | | | | | |
| 7-21 | | | | | | | | | | | | △ | △ | | | | | | | | | |
| 7-21 I | | | | | | | | | | | | △ | △ | | | | | | | | | |
| 7-17Z | | | | | | | | | △ | △ | | | | | | | | | | | | |
| 7-12Z | | | | | | | △ | △ | | | | | | | | | | | | | | |
| 7-22 | △ | | | | | | | | | | | | | | | | | | | | | |
| 7-27 | | | | | | | | | △ | | △ | | | | | | | | | | △ | |
| 7-27 I | | | | | | | | | | △ | △ | | | △ | | | | | | | △ | |
| 7-24 | | △ | △ | | | | | | | | | | | △ | | | | △ | | | | |
| 7-24 I | | | △ | △ | | | | | | | | | | | | | △ | △ | | | | |
| 7-28 | | | | △ | | | | | | | | | | | △ | | | | | | △ | |
| 7-28 I | | | | | △ | | | | | | | | | | | △ | | | | | △ | |
| 7-31 | △ | | | | | | △ | | | | | | | | △ | | | | | | | △ |
| 7-31 I | | | | | | | | △ | | | | | | | △ | | | | | | | △ |
| 7-25 | | | | △ | | △ | △ | | | | | | | | △ | | | | | | △ | |
| 7-25 I | | | △ | | | △ | | △ | | | | | | | | | △ | | | | △ | |
| 7-26 | | | | | | | | | | | | | | | | | △ | △ | | | | |
| 7-26 I | | | | | | | | | | | | | | | △ | | △ | | | | | |
| 7-30 | | | △ | | | | | | | | | | △ | | | | △ | | | | | |
| 7-30 I | | △ | | | | | | | | | | △ | | | | | △ | | | | | |
| 7-23 | | | | | | △ | | | | △ | | | △ | | | | | △ | | | | |
| 7-23 I | | | | | | △ | | | △ | | | | △ | | | | | △ | | | | |
| 7-29 | | | △ | | | | △ | | △ | | | | | | | | △ | △ | | | | |
| 7-29 I | | △ | | | | | | △ | | △ | | | | | △ | | | △ | | | | |
| 7-32 | △ | | | | | | | | | | △ | | | | △ | | | | | | △ | |
| 7-32 I | | | | | | | | | | | △ | | | | | △ | | | | | △ | |
| 7-33 | | | | | | | | | | | | | | | | | | △ | | △ | | △ |
| 7-34 | | | | | | | | | | | | | △ | △ | | | △ | △ | | △ | | |
| 7-35 | | | | | | | | | | | | | △ | △ | | | △ | | | △ | | |

Enrichment and reduction between $\mathscr{T}^8$ and $\mathscr{T}^9$

| | 9-12 | 9-10 | 9-09 | 9-06 | 9-11 I | 9-11 | 9-08 I | 9-08 | 9-07 I | 9-07 | 9-05 I | 9-05 | 9-03 I | 9-03 | 9-02 I | 9-02 | 9-04 I | 9-04 | 9-01 |
|---|---|---|---|---|---|---|---|---|---|---|---|---|---|---|---|---|---|---|---|
| 8-01 | | | | ◁ | | | | | | | | | | | ◁ | ◁ | | | ◁ |
| 8-02 | | | | | | | | | | | | | | ◁ | | ◁ | | | ◁ |
| 8-02 I | | | | | | | | | | | | | ◁ | | ◁ | | | ◁ | ◁ |
| 8-05 | | | | | | | | ◁ | | | ◁ | ◁ | | | | | ◁ | | ◁ |
| 8-05 I | | | | | | | ◁ | | | | | | | | | | | ◁ | ◁ |
| 8-04 | | | | | | | | | | ◁ | ◁ | ◁ | ◁ | ◁ | | | ◁ | | ◁ |
| 8-04 I | | | | | | | | | ◁ | | | | | | | | | | ◁ |
| 8-06 | | | ◁ | | | | | | | | | | | | | | | | |
| 8-03 | | | | ◁ | | | | | | | ◁ | ◁ | ◁ | ◁ | ◁ | ◁ | | | |
| 8-08 | | | | | | | | | | | | | | | | | ◁ | ◁ | |
| 8-07 | | | | | | | | | ◁ | | ◁ | ◁ | ◁ | ◁ | | | ◁ | ◁ | |
| 8-09 | | ◁ | | | | | | | | | | | | | | | | | |
| 8-12 | | | | | | | ◁ | | ◁ | ◁ | ◁ | ◁ | ◁ | ◁ | ◁ | ◁ | | | |
| 8-12 I | | | | | | | ◁ | | | | | | | ◁ | | | | | |
| 8-15Z | | | | | | | | | | | | | | | | | | | |
| 8-15Z I | | | | | | | | | | | | | | | | | | | |
| 8-11 | | | | ◁ | | | | ◁ | ◁ | ◁ | | | ◁ | ◁ | ◁ | ◁ | ◁ | ◁ | |
| 8-11 I | | | | ◁ | | | | | | | | | | | | | | | |
| 8-19 | ◁ | | | | ◁ | ◁ | | | | ◁ | | | | | | | | ◁ | |
| 8-19 I | ◁ | | | | ◁ | | | | | ◁ | | | | | | | | | |
| 8-18 | | ◁ | | | | | | | | | ◁ | ◁ | ◁ | ◁ | | | | ◁ | |
| 8-18 I | | ◁ | | | ◁ | ◁ | | | | | ◁ | | ◁ | | | | | | |
| 8-16 | | | ◁ | | | | ◁ | ◁ | | | ◁ | ◁ | | | | | ◁ | | |
| 8-16 I | | | ◁ | | | | | ◁ | | | | | | | | | | | |
| 8-29Z | | | ◁ | | | | | | | | | ◁ | | | ◁ | ◁ | | | |
| 8-29Z I | | | | | | | | | | | | | | | ◁ | ◁ | | | |
| 8-10 | | | | | | ◁ | ◁ | ◁ | ◁ | ◁ | | | | | | | | | |
| 8-13 | | ◁ | | | | | | | | | ◁ | ◁ | | | ◁ | ◁ | ◁ | | |
| 8-13 I | | ◁ | | | | | | | | | | | | | | | | | |
| 8-14 | | | | | | | | | | | | | | | ◁ | | ◁ | ◁ | |
| 8-14 I | | | | | | | | | | | | | | | | ◁ | | ◁ | |
| 8-20 | | | | | | | | | | | | | | ◁ | ◁ | | | | |
| 8-17 | | | ◁ | | ◁ | ◁ | | | ◁ | ◁ | | | | | | | | | |
| 8-22 | | | ◁ | ◁ | ◁ | ◁ | | ◁ | | | | | ◁ | | | | | | |
| 8-22 I | | | ◁ | ◁ | ◁ | ◁ | ◁ | ◁ | ◁ | ◁ | | | | | | | | | |
| 8-27 | | ◁ | | ◁ | ◁ | ◁ | ◁ | | ◁ | | | | | | | | | | |
| 8-27 I | | ◁ | | | | | | | | ◁ | | | | | | | | | |
| 8-24 | ◁ | | ◁ | | | | ◁ | | ◁ | ◁ | | | | | | | | | |
| 8-23 | | | | ◁ | | | | ◁ | | | | | | | | | | | |
| 8-21 | | | | | ◁ | | | ◁ | | | | | | | | | | | |
| 8-26 | | | | | | | ◁ | | | | | | | | | | | | |
| 8-25 | | | | | | | | | | | | | | | | | | | |
| 8-28 | | ◁ | | | | | | | | | | | | | | | | | |

# Appendix D. Table of eigenvalues

A primitive root of unity has the form:
$$\varepsilon = e^{\frac{2\pi i}{n}} = \cos\left(\frac{2\pi}{n}\right) + i\sin\left(\frac{2\pi}{n}\right)$$

## Case $n = 3$

The eigenvalues $\lambda_k$ of the circulant matrix $M_C$:

$$\varepsilon = e^{\frac{2\pi i}{3}} = \cos\left(\frac{2\pi}{3}\right) + i\sin\left(\frac{2\pi}{3}\right) = -\frac{1}{2} + i\frac{\sqrt{3}}{2}$$

$$\lambda_0 = x_1 + x_2 + x_3 = 12$$

$$\lambda_1 = x_1 + x_2\varepsilon + x_3\varepsilon^2 = x_1 - \frac{1}{2}(x_2 + x_3) + i\frac{\sqrt{3}}{2}(x_2 - x_3)$$

$$= \alpha + i\beta$$

$$\lambda_2 = \overline{\lambda_1}$$

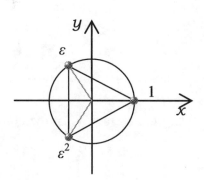

The eigenvalues $\lambda_k$ of the Hankel matrix $M_H$:

$$\lambda_0 = 12, \quad \lambda_{1,2}^H = \pm\sqrt{\lambda_1^C \cdot \overline{\lambda_1^C}}$$

| Prime form | Forte Name | $\lambda_k$ of $M_C$ $\lambda_0$ | $\alpha$ | $\beta$ | $\lambda_k$ of $M_H$ $\lambda_0$ | $\lambda_1, \lambda_2 = -\lambda_1$ |
|---|---|---|---|---|---|---|
| (1,1,10) | 3-01 | 12 | $-9/2$ | $-9\sqrt{3}/2$ | 12 | 9 |
| (1,2,9) | 3-02 | 12 | $-9/2$ | $-7\sqrt{3}/2$ | 12 | $\sqrt{57}$ |
| (2,1,9) | 3-02 I | 12 | $-3$ | $-4\sqrt{3}$ | 12 | $\sqrt{57}$ |
| (1,3,8) | 3-03 | 12 | $-9/2$ | $-5\sqrt{3}/2$ | 12 | $\sqrt{39}$ |
| (3,1,8) | 3-03 I | 12 | $-3/2$ | $-7\sqrt{3}/2$ | 12 | $\sqrt{39}$ |
| (2,2,8) | 3-06 | 12 | $-3$ | $-3\sqrt{3}$ | 12 | 6 |
| (1,4,7) | 3-04 | 12 | $-9/2$ | $-3\sqrt{3}/2$ | 12 | $3\sqrt{3}$ |
| (4,1,7) | 3-04 I | 12 | 0 | $-3\sqrt{3}$ | 12 | $3\sqrt{3}$ |
| (1,5,6) | 3-05 | 12 | $-9/2$ | $-\sqrt{3}/2$ | 12 | $\sqrt{21}$ |
| (5,1,6) | 3-05 I | 12 | $3/2$ | $-5\sqrt{3}/2$ | 12 | $\sqrt{21}$ |
| (2,3,7) | 3-07 | 12 | $-3$ | $-2\sqrt{3}$ | 12 | $\sqrt{21}$ |
| (3,2,7) | 3-07 I | 12 | $-3/2$ | $-5\sqrt{3}/2$ | 12 | $\sqrt{21}$ |
| (2,4,6) | 3-08 | 12 | $-3$ | $-\sqrt{3}$ | 12 | $2\sqrt{3}$ |
| (4,2,6) | 3-08 I | 12 | 0 | $-2\sqrt{3}$ | 12 | $2\sqrt{3}$ |
| (2,5,5) | 3-09 | 12 | $-3$ | 0 | 12 | 3 |
| (3,3,6) | 3-10 | 12 | $-3/2$ | $-3\sqrt{3}/2$ | 12 | 3 |
| (3,4,5) | 3-11 | 12 | $-3/2$ | $-\sqrt{3}/2$ | 12 | $\sqrt{3}$ |
| (4,3,5) | 3-11 I | 12 | 0 | $-\sqrt{3}$ | 12 | $\sqrt{3}$ |
| (4,4,4) | 3-12 | 12 | 0 | 0 | 12 | 0 |

**Case $n = 4$**

The eigenvalues $\lambda_k$ of the circulant matrix $M_C$:

$$\varepsilon = e^{\frac{2\pi i}{4}} = \cos\left(\frac{\pi}{2}\right) + i\sin\left(\frac{\pi}{2}\right) = i$$

$$\lambda_0 = x_1 + x_2 + x_3 + x_4 = 12$$

$$\lambda_1 = x_1 + x_2\varepsilon + x_3\varepsilon^2 + x_4\varepsilon^3 = x_1 + ix_2 - x_3 - ix_4 = (x_1 - x_3) + i(x_2 - x_4)$$
$$= \alpha + i\beta$$

$$\lambda_2 = x_1 + x_2\varepsilon^2 + x_3\varepsilon^4 + x_4\varepsilon^6 = x_1 - x_2 + x_3 - x_4$$

$$\lambda_3 = x_1 + x_2\varepsilon^3 + x_3\varepsilon^6 + x_4\varepsilon^9 = \overline{\lambda_1}$$

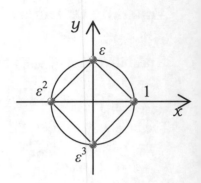

The eigenvalues $\lambda_k$ of the Hankel matrix $M_H$:

$$\lambda_0 = 12, \quad \lambda_{1,3}^H = \pm\sqrt{\lambda_1^C \cdot \overline{\lambda_1^C}}, \quad \lambda_2^H = \lambda_2^C$$

| Prime form | Forte Name | $\lambda_k$ of $M_C$ | | | | $\lambda_k$ of $M_H$ | | |
|---|---|---|---|---|---|---|---|---|
| | | $\lambda_0$ | $\alpha$ | $\beta$ | $\lambda_2$ | $\lambda_0$ | $\lambda_1, \lambda_3 = -\lambda_1$ | $\lambda_2$ |
| (1,1,1,9) | 4-01 | 12 | 0 | $-8$ | $-8$ | 12 | 8 | $-8$ |
| (1,1,2,8) | 4-02 | 12 | $-1$ | $-7$ | $-6$ | 12 | $5\sqrt{2}$ | $-6$ |
| (2,1,1,8) | 4-02 I | 12 | 1 | $-7$ | $-6$ | 12 | $5\sqrt{2}$ | $-6$ |
| (1,2,1,8) | 4-03 | 12 | 0 | $-6$ | $-8$ | 12 | 6 | $-8$ |
| (1,1,3,7) | 4-04 | 12 | $-2$ | $-6$ | $-4$ | 12 | $2\sqrt{10}$ | $-4$ |
| (3,1,1,7) | 4-04 I | 12 | 2 | $-6$ | $-4$ | 12 | $2\sqrt{10}$ | $-4$ |
| (1,3,1,7) | 4-07 | 12 | 0 | $-4$ | $-8$ | 12 | 4 | $-8$ |
| (1,2,2,7) | 4-11 | 12 | $-1$ | $-5$ | $-6$ | 12 | $\sqrt{26}$ | $-6$ |
| (2,2,1,7) | 4-11 I | 12 | 1 | $-5$ | $-6$ | 12 | $\sqrt{26}$ | $-6$ |
| (2,1,2,7) | 4-10 | 12 | 0 | $-6$ | $-4$ | 12 | 6 | $-4$ |
| (1,1,4,6) | 4-05 | 12 | $-3$ | $-5$ | $-2$ | 12 | $\sqrt{34}$ | $-2$ |
| (4,1,1,6) | 4-05 I | 12 | 3 | $-5$ | $-2$ | 12 | $\sqrt{34}$ | $-2$ |
| (1,4,1,6) | 4-08 | 12 | 0 | $-2$ | $-8$ | 12 | 2 | $-8$ |
| (1,1,5,5) | 4-06 | 12 | $-4$ | $-4$ | 0 | 12 | $4\sqrt{2}$ | 0 |
| (1,5,1,5) | 4-09 | 12 | 0 | 0 | $-8$ | 12 | 0 | $-8$ |
| (1,2,3,6) | 4-13 | 12 | $-2$ | $-4$ | $-4$ | 12 | $2\sqrt{5}$ | $-4$ |
| (3,2,1,6) | 4-13 I | 12 | 2 | $-4$ | $-4$ | 12 | $2\sqrt{5}$ | $-4$ |
| (1,3,2,6) | 4-15Z | 12 | $-1$ | $-3$ | $-6$ | 12 | $\sqrt{10}$ | $-6$ |
| (2,3,1,6) | 4-15Z I | 12 | 1 | $-3$ | $-6$ | 12 | $\sqrt{10}$ | $-6$ |
| (2,1,3,6) | 4-12 | 12 | $-1$ | $-5$ | $-2$ | 12 | $\sqrt{26}$ | $-2$ |
| (3,1,2,6) | 4-12 I | 12 | 1 | $-5$ | $-2$ | 12 | $\sqrt{26}$ | $-2$ |
| (2,2,2,6) | 4-21 | 12 | 0 | $-4$ | $-4$ | 12 | 4 | $-4$ |
| (1,2,4,5) | 4-29Z | 12 | $-3$ | $-3$ | $-2$ | 12 | $3\sqrt{2}$ | $-2$ |
| (4,2,1,5) | 4-29Z I | 12 | 3 | $-3$ | $-2$ | 12 | $3\sqrt{2}$ | $-2$ |

| (1,4,2,5) | 4-16 | 12 | −1 | −1 | −6 | 12 | $\sqrt{2}$ | −6 |
|---|---|---|---|---|---|---|---|---|
| (2,4,1,5) | 4-16 I | 12 | 1 | −1 | −6 | 12 | $\sqrt{2}$ | −6 |
| (2,1,4,5) | 4-14 | 12 | −2 | −4 | 0 | 12 | $2\sqrt{5}$ | 0 |
| (4,1,2,5) | 4-14 I | 12 | 2 | −4 | 0 | 12 | $2\sqrt{5}$ | 0 |
| (1,3,3,5) | 4-18 | 12 | −2 | −2 | −4 | 12 | $2\sqrt{2}$ | −4 |
| (3,3,1,5) | 4-18 I | 12 | 2 | −2 | −4 | 12 | $2\sqrt{2}$ | −4 |
| (3,1,3,5) | 4-17 | 12 | 0 | −4 | 0 | 12 | 4 | 0 |
| (1,3,4,4) | 4-19 | 12 | −3 | −1 | −2 | 12 | $\sqrt{10}$ | −2 |
| (3,1,4,4) | 4-19 I | 12 | −1 | −3 | 2 | 12 | $\sqrt{10}$ | 2 |
| (2,2,3,5) | 4-22 | 12 | −1 | −3 | −2 | 12 | $\sqrt{10}$ | −2 |
| (3,2,2,5) | 4-22 I | 12 | 1 | −3 | −2 | 12 | $\sqrt{10}$ | −2 |
| (2,3,2,5) | 4-23 | 12 | 0 | −2 | −4 | 12 | 2 | −4 |
| (1,4,3,4) | 4-20 | 12 | −2 | 0 | −4 | 12 | 2 | −4 |
| (2,2,4,4) | 4-24 | 12 | −2 | −2 | 0 | 12 | $2\sqrt{2}$ | 0 |
| (2,4,2,4) | 4-25 | 12 | 0 | 0 | −4 | 12 | 0 | −4 |
| (2,3,3,4) | 4-27 | 12 | −1 | −1 | −2 | 12 | $\sqrt{2}$ | −2 |
| (3,3,2,4) | 4-27 I | 12 | 1 | −1 | −2 | 12 | $\sqrt{2}$ | −2 |
| (3,2,3,4) | 4-26 | 12 | 0 | −2 | 0 | 12 | 2 | 0 |
| (3,3,3,3) | 4-28 | 12 | 0 | 0 | 0 | 12 | 0 | 0 |

## Case $n = 5$

The eigenvalues $\lambda_k$ of the circulant matrix $M_C$:

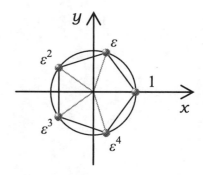

$$\varepsilon = e^{\frac{2\pi i}{5}} = \cos\left(\frac{2\pi}{5}\right) + i\sin\left(\frac{2\pi}{5}\right) = \frac{-1+\sqrt{5}}{4} + \frac{\sqrt{10+2\sqrt{5}}}{4}i$$

$$\lambda_0 = x_1 + x_2 + x_3 + x_4 + x_5 = 12$$

$$\lambda_1 = x_1 + x_2\varepsilon + x_3\varepsilon^2 + x_4\varepsilon^3 + x_5\varepsilon^4$$

$$= x_1 + \frac{\left(-1+\sqrt{5}\right)(x_2+x_5) - \left(1+\sqrt{5}\right)(x_3+x_4)}{4} + i\left(\frac{\left(\sqrt{10+2\sqrt{5}}\right)(x_2-x_5) + \left(\sqrt{10-2\sqrt{5}}\right)(x_3-x_4)}{4}\right)$$

$$\lambda_2 = x_1 + x_2\varepsilon^2 + x_3\varepsilon^4 + x_4\varepsilon^6 + x_5\varepsilon^8 = x_1 + x_2\varepsilon^2 + x_3\varepsilon^4 + x_4\varepsilon + x_5\varepsilon^3 = x_1 + x_4\varepsilon + x_2\varepsilon^2 + x_5\varepsilon^3 + x_3\varepsilon^4$$

$$= x_1 + \frac{\left(-1+\sqrt{5}\right)(x_4+x_3) - \left(1+\sqrt{5}\right)(x_2+x_5)}{4} + i\left(\frac{\left(\sqrt{10+2\sqrt{5}}\right)(x_4-x_3) + \left(\sqrt{10-2\sqrt{5}}\right)(x_2-x_5)}{4}\right)$$

**Case $n = 6$**

The eigenvalues $\lambda_k$ of the circulant matrix $M_C$:

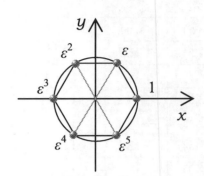

$$\varepsilon = e^{\frac{2\pi i}{6}} = \cos\left(\frac{\pi}{3}\right) + i\sin\left(\frac{\pi}{3}\right) \qquad = \frac{1}{2} + \frac{\sqrt{3}}{2}i$$

$$\lambda_0 = x_1 + x_2 + x_3 + x_4 + x_5 + x_6 = 12$$

$$\lambda_1 = x_1 + x_2\varepsilon + x_3\varepsilon^2 + x_4\varepsilon^3 + x_5\varepsilon^4 + x_6\varepsilon^5$$

$$= x_1 + \left(\frac{1}{2} + \frac{\sqrt{3}}{2}i\right)x_2 + \left(-\frac{1}{2} + \frac{\sqrt{3}}{2}i\right)x_3 - x_4 - \left(\frac{1}{2} + \frac{\sqrt{3}}{2}i\right)x_5 + \left(\frac{1}{2} - \frac{\sqrt{3}}{2}i\right)x_6$$

$$= x_1 - x_4 + \frac{1}{2}(x_2 + x_6 - x_3 - x_5) + i\frac{\sqrt{3}}{2}(x_2 + x_3 - x_5 - x_6)$$

$$= \alpha_1 + i\beta_1$$

$$\lambda_2 = x_1 + x_2\varepsilon^2 + x_3\varepsilon^4 + x_4\varepsilon^6 + x_5\varepsilon^8 + x_6\varepsilon^{10} = (x_1 + x_4) + (x_2 + x_5)\varepsilon^2 + (x_3 + x_6)\varepsilon^4$$

$$= (x_1 + x_4) + (x_2 + x_5)\cdot\left(-\frac{1}{2} + \frac{\sqrt{3}}{2}i\right) - (x_3 + x_6)\cdot\left(\frac{1}{2} + \frac{\sqrt{3}}{2}i\right)$$

$$= x_1 + x_4 - \frac{1}{2}(x_2 + x_5 + x_3 + x_6) + i\frac{\sqrt{3}}{2}(x_2 + x_5 - x_3 - x_6)$$

$$= \alpha_2 + i\beta_2$$

$$\lambda_3 = x_1 - x_2 + x_3 - x_4 + x_5 - x_6$$

$$\lambda_4 = \overline{\lambda_2}, \quad \lambda_5 = \overline{\lambda_1}$$

The eigenvalues $\lambda_k$ of the Hankel matrix $M_H$:

$$\lambda_0 = 12, \quad \lambda_{1,5}^H = \pm\sqrt{\lambda_1^C \cdot \overline{\lambda_1^C}}, \quad \lambda_{2,4}^H = \pm\sqrt{\lambda_2^C \cdot \overline{\lambda_2^C}}, \quad \lambda_3^H = \lambda_3^C$$

| Prime form | Forte Name | $\lambda_k$ of $M_C$ | | | | | | $\lambda_k$ of $M_H$ | | | |
|---|---|---|---|---|---|---|---|---|---|---|---|
| | | $\lambda_0$ | $\alpha_1$ | $\beta_1$ | $\alpha_2$ | $\beta_2$ | $\lambda_3$ | $\lambda_0$ | $\lambda_1$ | $\lambda_2$ | $\lambda_3$ |
| (1,1,1,1,1,7) | 6-01 | 12 | 3 | $-6\sqrt{3}/2$ | $-3$ | $-3\sqrt{3}$ | $-6$ | 12 | 6 | 6 | $-6$ |
| (1,1,1,2,1,6) | 6-03Z | 12 | 3/2 | $-5\sqrt{3}/2$ | $-3/2$ | $-5\sqrt{3}/2$ | $-6$ | 12 | $\sqrt{21}$ | $\sqrt{21}$ | $-6$ |
| (1,2,1,1,1,6) | 6-03Z I | 12 | 3 | $-2\sqrt{3}$ | $-3$ | $-2\sqrt{3}$ | $-6$ | 12 | $\sqrt{21}$ | $\sqrt{21}$ | $-6$ |
| (1,1,1,1,2,6) | 6-02 | 12 | 2 | $-3\sqrt{3}$ | $-3$ | $-2\sqrt{3}$ | $-4$ | 12 | $\sqrt{31}$ | $\sqrt{21}$ | $-4$ |
| (2,1,1,1,1,6) | 6-02 I | 12 | 7/2 | $-5\sqrt{3}/2$ | $-3/2$ | $-5\sqrt{3}/2$ | $-4$ | 12 | $\sqrt{31}$ | $\sqrt{21}$ | $-4$ |
| (1,1,2,1,1,6) | 6-04Z | 12 | 2 | $-2\sqrt{3}$ | $-3$ | $-3\sqrt{3}$ | $-4$ | 12 | 4 | 6 | $-4$ |
| (1,1,1,3,1,5) | 6-05 | 12 | 0 | $-2\sqrt{3}$ | 0 | $-2\sqrt{3}$ | $-6$ | 12 | $2\sqrt{3}$ | $2\sqrt{3}$ | $-6$ |
| (1,3,1,1,1,5) | 6-05 I | 12 | 3 | $-\sqrt{3}$ | $-3$ | $-\sqrt{3}$ | $-6$ | 12 | $2\sqrt{3}$ | $2\sqrt{3}$ | $-6$ |
| (1,1,1,1,3,5) | 6-36Z | 12 | 1 | $-3\sqrt{3}$ | $-3$ | $-\sqrt{3}$ | $-2$ | 12 | $2\sqrt{7}$ | $2\sqrt{3}$ | $-2$ |
| (3,1,1,1,1,5) | 6-36Z I | 12 | 4 | $-2\sqrt{3}$ | 0 | $-2\sqrt{3}$ | $-2$ | 12 | $2\sqrt{7}$ | $2\sqrt{3}$ | $-2$ |
| (1,1,3,1,1,5) | 6-06Z | 12 | 1 | $-\sqrt{3}$ | $-3$ | $-3\sqrt{3}$ | $-2$ | 12 | 2 | 6 | $-2$ |
| (1,2,1,1,2,5) | 6-10Z | 12 | 1 | $-2\sqrt{3}$ | $-3$ | $-\sqrt{3}$ | $-4$ | 12 | 4 | $2\sqrt{3}$ | $-4$ |
| (2,1,1,2,1,5) | 6-10Z I | 12 | 1 | $-2\sqrt{3}$ | 0 | $-2\sqrt{3}$ | $-4$ | 12 | 4 | $2\sqrt{3}$ | $-4$ |
| (1,1,1,2,2,5) | 6-09 | 12 | 1/2 | $-5\sqrt{3}/2$ | $-3/2$ | $-3\sqrt{3}/2$ | $-4$ | 12 | $\sqrt{19}$ | 3 | $-4$ |
| (2,2,1,1,1,5) | 6-09 I | 12 | 7/2 | $-3\sqrt{3}/2$ | $-3/2$ | $-3\sqrt{3}/2$ | $-4$ | 12 | $\sqrt{19}$ | 3 | $-4$ |
| (1,1,2,2,1,5) | 6-12Z | 12 | 1/2 | $-3\sqrt{3}/2$ | $-3/2$ | $-5\sqrt{3}/2$ | $-4$ | 12 | $\sqrt{7}$ | $\sqrt{21}$ | $-4$ |
| (1,2,2,1,1,5) | 6-12Z I | 12 | 2 | $-\sqrt{3}$ | $-3$ | $-2\sqrt{3}$ | $-4$ | 12 | $\sqrt{7}$ | $\sqrt{21}$ | $-4$ |
| (1,1,2,1,2,5) | 6-11Z | 12 | 1 | $-2\sqrt{3}$ | $-3$ | $-2\sqrt{3}$ | $-2$ | 12 | $\sqrt{13}$ | $\sqrt{21}$ | $-2$ |
| (2,1,2,1,1,5) | 6-11Z I | 12 | 5/2 | $-3\sqrt{3}/2$ | $-3/2$ | $-5\sqrt{3}/2$ | $-2$ | 12 | $\sqrt{13}$ | $\sqrt{21}$ | $-2$ |
| (1,2,1,2,1,5) | 6-13Z | 12 | 3/2 | $-3\sqrt{3}/2$ | $-3/2$ | $-3\sqrt{3}/2$ | $-6$ | 12 | 3 | 3 | $-6$ |
| (1,1,1,4,1,4) | 6-38Z | 12 | $-3/2$ | $-3\sqrt{3}/2$ | $3/2$ | $-3\sqrt{3}/2$ | $-6$ | 12 | 3 | 3 | $-6$ |
| (2,1,1,1,2,5) | 6-08 | 12 | 5/2 | $-5\sqrt{3}/2$ | $-3/2$ | $-3\sqrt{3}/2$ | $-2$ | 12 | 5 | 3 | $-2$ |
| (1,1,1,1,4,4) | 6-37Z | 12 | 0 | $-3\sqrt{3}$ | $-3$ | 0 | 0 | 12 | $3\sqrt{3}$ | 3 | 0 |
| (1,1,4,1,1,4) | 6-07 | 12 | 0 | 0 | $-3$ | $-3\sqrt{3}$ | 0 | 12 | 0 | 6 | 0 |
| (1,3,1,1,2,4) | 6-16 | 12 | 2 | $-\sqrt{3}$ | $-3$ | 0 | $-4$ | 12 | 7 | 3 | $-4$ |
| (2,1,1,3,1,4) | 6-16 I | 12 | 1/2 | $-3\sqrt{3}/2$ | $3/2$ | $-3\sqrt{3}/2$ | $-4$ | 12 | 7 | 3 | $-4$ |
| (1,2,1,1,3,4) | 6-14 | 12 | 1 | $-2\sqrt{3}$ | $-3$ | 0 | $-2$ | 12 | $\sqrt{13}$ | 3 | $-2$ |
| (3,1,1,2,1,4) | 6-14 I | 12 | 5/2 | $-3\sqrt{3}/2$ | $3/2$ | $-3\sqrt{3}/2$ | $-2$ | 12 | $\sqrt{13}$ | 3 | $-2$ |
| (1,1,2,3,1,4) | 6-17Z | 12 | $-1$ | $-\sqrt{3}$ | 0 | $-2\sqrt{3}$ | $-4$ | 12 | 2 | $2\sqrt{3}$ | $-4$ |
| (1,3,2,1,1,4) | 6-17Z I | 12 | 2 | 0 | $-3$ | $-\sqrt{3}$ | $-4$ | 12 | 2 | $2\sqrt{3}$ | $-4$ |
| (1,1,1,3,2,4) | 6-41Z | 12 | $-1$ | $-2\sqrt{3}$ | 0 | $-\sqrt{3}$ | $-4$ | 12 | $\sqrt{13}$ | $\sqrt{3}$ | $-4$ |
| (2,3,1,1,1,4) | 6-41Z I | 12 | 7/2 | $-\sqrt{3}/2$ | $-3/2$ | $-\sqrt{3}/2$ | $-4$ | 12 | $\sqrt{13}$ | $\sqrt{3}$ | $-4$ |
| (1,1,1,2,3,4) | 6-40Z | 12 | $-1/2$ | $-5\sqrt{3}/2$ | $-3/2$ | $-\sqrt{3}/2$ | $-2$ | 12 | $\sqrt{19}$ | $\sqrt{3}$ | $-2$ |
| (3,2,1,1,1,4) | 6-40Z I | 12 | 4 | $-\sqrt{3}$ | 0 | $-\sqrt{3}$ | $-2$ | 12 | $\sqrt{19}$ | $\sqrt{3}$ | $-2$ |
| (1,2,1,3,1,4) | 6-19Z | 12 | 0 | $-\sqrt{3}$ | 0 | $-\sqrt{3}$ | $-6$ | 12 | $\sqrt{3}$ | $\sqrt{3}$ | $-6$ |
| (1,3,1,2,1,4) | 6-19Z I | 12 | 3/2 | $-\sqrt{3}/2$ | $-3/2$ | $-\sqrt{3}/2$ | $-6$ | 12 | $\sqrt{3}$ | $\sqrt{3}$ | $-6$ |
| (1,1,3,2,1,4) | 6-18 | 12 | $-1/2$ | $-\sqrt{3}/2$ | $-3/2$ | $-5\sqrt{3}/2$ | $-2$ | 12 | 1 | $\sqrt{21}$ | $-2$ |
| (1,2,3,1,1,4) | 6-18 I | 12 | 1 | 0 | $-3$ | $-2\sqrt{3}$ | $-2$ | 12 | 1 | $\sqrt{21}$ | $-2$ |
| (1,1,2,1,3,4) | 6-15 | 12 | 0 | $-2\sqrt{3}$ | $-3$ | $-\sqrt{3}/2$ | 0 | 12 | $2\sqrt{3}$ | $2\sqrt{3}$ | 0 |
| (3,1,2,1,1,4), | 6-15 I | 12 | 3 | $-\sqrt{3}$ | 0 | $-2\sqrt{3}$ | 0 | 12 | $2\sqrt{3}$ | $2\sqrt{3}$ | 0 |

| Prime form | Forte Name | $\lambda_k$ of $M_C$ | | | | | | $\lambda_k$ of $M_H$ | | | |
|---|---|---|---|---|---|---|---|---|---|---|---|
| | | $\lambda_0$ | $\alpha_1$ | $\beta_1$ | $\alpha_2$ | $\beta_2$ | $\lambda_3$ | $\lambda_0$ | $\lambda_1$ | $\lambda_2$ | $\lambda_3$ |
| (2,1,1,1,3,4) | 6-39Z | 12 | 3/2 | $-5\sqrt{3}/2$ | $-3/2$ | $-\sqrt{3}/2$ | 0 | 12 | $\sqrt{21}$ | $\sqrt{3}$ | 0 |
| (3,1,1,1,2,4) | 6-39Z I | 12 | 3 | $-2\sqrt{3}$ | 0 | $-\sqrt{3}$ | 0 | 12 | $\sqrt{21}$ | $\sqrt{3}$ | 0 |
| (1,1,3,1,2,4) | 6-43Z | 12 | 0 | $-\sqrt{3}$ | $-3$ | $-2\sqrt{3}$ | 0 | 12 | $\sqrt{3}$ | $\sqrt{21}$ | 0 |
| (2,1,3,1,1,4) | 6-43Z I | 12 | 3/2 | $-\sqrt{3}/2$ | $-3/2$ | $-5\sqrt{3}/2$ | 0 | 12 | $\sqrt{3}$ | $\sqrt{21}$ | 0 |
| (1,1,2,2,2,4) | 6-22 | 12 | $-1/2$ | $-3\sqrt{3}/2$ | $-3/2$ | $-3\sqrt{3}/2$ | $-2$ | 12 | $\sqrt{7}$ | 3 | $-2$ |
| (2,2,2,1,1,4) | 6-22 I | 12 | 5/2 | $-\sqrt{3}/2$ | $-3/2$ | $-3\sqrt{3}/2$ | $-2$ | 12 | $\sqrt{7}$ | 3 | $-2$ |
| (1,2,2,1,2,4) | 6-25Z | 12 | 1 | $-\sqrt{3}$ | $-3$ | $-\sqrt{3}$ | $-2$ | 12 | 2 | $2\sqrt{3}$ | $-2$ |
| (2,1,2,2,1,4) | 6-25Z I | 12 | 1 | $-\sqrt{3}$ | 0 | $-2\sqrt{3}$ | $-2$ | 12 | 2 | $2\sqrt{3}$ | $-2$ |
| (1,1,3,1,3,3) | 6-44Z | 12 | $-1$ | $-\sqrt{3}$ | $-3$ | $-\sqrt{3}$ | 2 | 12 | 2 | $2\sqrt{3}$ | 2 |
| (1,3,1,1,3,3) | 6-44Z I | 12 | 1 | $-\sqrt{3}$ | $-3$ | $\sqrt{3}$ | $-2$ | 12 | 2 | $2\sqrt{3}$ | $-2$ |
| (1,2,1,2,2,4) | 6-24Z | 12 | 1/2 | $-3\sqrt{3}/2$ | $-3/2$ | $-\sqrt{3}/2$ | $-4$ | 12 | $\sqrt{7}$ | $\sqrt{3}$ | $-4$ |
| (2,2,1,2,1,4) | 6-24Z I | 12 | 2 | $-\sqrt{3}$ | 0 | $-\sqrt{3}$ | $-4$ | 12 | $\sqrt{7}$ | $\sqrt{3}$ | $-4$ |
| (2,1,1,2,2,4) | 6-21 | 12 | 1 | $-2\sqrt{3}$ | 0 | $-\sqrt{3}$ | $-2$ | 12 | $\sqrt{13}$ | $\sqrt{3}$ | $-2$ |
| (2,2,1,1,2,4) | 6-21 I | 12 | 5/2 | $-3\sqrt{3}/2$ | $-3/2$ | $-\sqrt{3}/2$ | $-2$ | 12 | $\sqrt{13}$ | $\sqrt{3}$ | $-2$ |
| (1,2,2,2,1,4) | 6-26Z | 12 | 1/2 | $-\sqrt{3}/2$ | $-3/2$ | $-3\sqrt{3}/2$ | $-4$ | 12 | 1 | 3 | $-4$ |
| (2,1,2,1,2,4) | 6-23Z | 12 | 3/2 | $-3$ | $-3/2$ | $-3\sqrt{3}/2$ | 0 | 12 | 3 | 3 | 0 |
| (1,1,1,3,3,3) | 6-42Z | 12 | $-2$ | $-2\sqrt{3}$ | 0 | 0 | $-2$ | 12 | 4 | 0 | $-2$ |
| (1,3,1,3,1,3) | 6-20 | 12 | 0 | 0 | 0 | 0 | $-6$ | 12 | 0 | 0 | $-6$ |
| (1,2,1,2,3,3) | 6-27 | 12 | $-1/2$ | $-3\sqrt{3}/2$ | $-3/2$ | $\sqrt{3}/2$ | $-2$ | 12 | $\sqrt{7}$ | $\sqrt{3}$ | $-2$ |
| (2,1,2,1,3,3) | 6-27 I | 12 | 1/2 | $-3\sqrt{3}/2$ | $-3/2$ | $-\sqrt{3}/2$ | 2 | 12 | $\sqrt{7}$ | $\sqrt{3}$ | 2 |
| (1,1,2,3,2,3) | 6-47Z | 12 | $-2$ | $-\sqrt{3}$ | 0 | $-\sqrt{3}$ | $-2$ | 12 | $\sqrt{7}$ | $\sqrt{3}$ | $-2$ |
| (2,1,1,3,2,3) | 6-47Z I | 12 | $-1/2$ | $-3\sqrt{3}/2$ | 3/2 | $-\sqrt{3}/2$ | $-2$ | 12 | $\sqrt{7}$ | $\sqrt{3}$ | $-2$ |
| (2,1,3,1,2,3) | 6-29Z | 12 | 1/2 | $-\sqrt{3}/2$ | $-3/2$ | $-3\sqrt{3}/2$ | 2 | 12 | 1 | 3 | 2 |
| (1,3,2,1,2,3) | 6-50Z | 12 | 1 | 0 | $-3$ | 0 | $-2$ | 12 | 1 | 3 | $-2$ |
| (1,3,1,2,2,3) | 6-31 | 12 | 1/2 | $-\sqrt{3}/2$ | $-3/2$ | $\sqrt{3}/2$ | $-4$ | 12 | 1 | $\sqrt{3}$ | $-4$ |
| (2,2,1,3,1,3) | 6-31 I | 12 | 1/2 | $-\sqrt{3}/2$ | 3/2 | $-\sqrt{3}/2$ | $-4$ | 12 | 1 | $\sqrt{3}$ | $-4$ |
| (1,2,2,1,3,3) | 6-28Z | 12 | 0 | $-\sqrt{3}$ | $-6$ | 0 | 0 | 12 | $\sqrt{3}$ | 3 | 0 |
| (1,1,2,2,3,3) | 6-46Z | 12 | $-3/2$ | $-3\sqrt{3}/2$ | $-3/2$ | $-\sqrt{3}/2$ | 0 | 12 | 3 | $\sqrt{3}$ | 0 |
| (2,2,1,1,3,3) | 6-46Z I | 12 | 3/2 | $-3\sqrt{3}/2$ | $-3/2$ | $\sqrt{3}/2$ | 0 | 12 | 3 | $\sqrt{3}$ | 0 |
| (1,1,3,2,2,3) | 6-48Z | 12 | $-3/2$ | $-\sqrt{3}/2$ | $-3/2$ | $-3\sqrt{3}/2$ | 0 | 12 | $\sqrt{3}$ | 3 | 0 |
| (1,2,1,3,2,3) | 6-49Z | 12 | $-1$ | $-\sqrt{3}$ | 0 | 0 | $-4$ | 12 | 2 | 0 | $-4$ |
| (1,2,3,1,2,3) | 6-30 | 12 | 0 | 0 | $-3$ | $-\sqrt{3}$ | 0 | 12 | 0 | $2\sqrt{3}$ | 0 |
| (2,1,3,2,1,3) | 6-30 I | 12 | 0 | 0 | 0 | $-2\sqrt{3}$ | 0 | 12 | 0 | $2\sqrt{3}$ | 0 |
| (2,1,1,2,3,3) | 6-45Z | 12 | 0 | $-2\sqrt{3}$ | 0 | 0 | 0 | 12 | $2\sqrt{3}$ | 0 | 0 |
| (1,2,2,2,2,3) | 6-34 | 12 | $-1/2$ | $-\sqrt{3}/2$ | $-3/2$ | $-\sqrt{3}/2$ | $-2$ | 12 | 1 | $\sqrt{3}$ | $-2$ |
| (2,2,2,2,1,3) | 6-34 I | 12 | 1 | 0 | 0 | $-\sqrt{3}$ | $-2$ | 12 | 1 | $\sqrt{3}$ | $-2$ |
| (2,1,2,2,2,3) | 6-33 | 12 | 0 | $-\sqrt{3}$ | 0 | $-\sqrt{3}$ | 0 | 12 | $\sqrt{3}$ | $\sqrt{3}$ | 0 |
| (2,2,2,1,2,3) | 6-33 I | 12 | 3/2 | $-\sqrt{3}/2$ | $-3/2$ | $-\sqrt{3}/2$ | 0 | 12 | $\sqrt{3}$ | $\sqrt{3}$ | 0 |
| (2,2,1,2,2,3) | 6-32 | 12 | 1 | $-\sqrt{3}$ | 0 | 0 | $-2$ | 12 | 2 | 0 | $-2$ |
| (2,2,2,2,2,2) | 6-35 | 12 | 0 | 0 | 0 | 0 | 0 | 12 | 0 | 0 | 0 |

**Case *n* = 7**

$$\varepsilon = e^{\frac{2\pi i}{7}} = \cos\left(\frac{2\pi}{7}\right) + i\sin\left(\frac{2\pi}{7}\right)$$

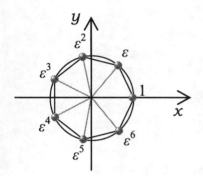

**Case *n* = 8**

$$\varepsilon = e^{\frac{2\pi i}{8}} = \cos\left(\frac{\pi}{4}\right) + i\sin\left(\frac{\pi}{4}\right)$$

$$= \frac{\sqrt{2}}{2} + \frac{\sqrt{2}}{2}\, i$$

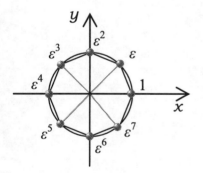

**Case *n* = 9**

$$\varepsilon = e^{\frac{2\pi i}{9}} = \cos\left(\frac{2\pi}{9}\right) + i\sin\left(\frac{2\pi}{9}\right)$$

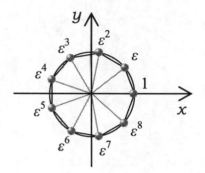

**Case *n* = 10**

$$\varepsilon = e^{\frac{2\pi i}{10}} = \cos\left(\frac{\pi}{5}\right) + i\sin\left(\frac{\pi}{5}\right)$$

$$= \frac{1+\sqrt{5}}{4} + \frac{\sqrt{10-2\sqrt{5}}}{4}\, i$$

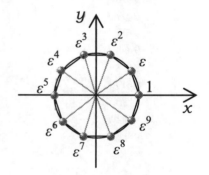

**Case *n* = 11**

$$\varepsilon = e^{\frac{2\pi i}{11}} = \cos\left(\frac{2\pi}{11}\right) + i\sin\left(\frac{2\pi}{11}\right)$$

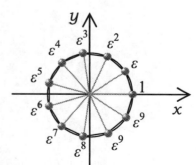

## Appendix E. Chord Complementation

| Forte Name | Complement |
|---|---|
| 2-01 | 10-01 |
| 2-02 | 10-02 |
| 2-03 | 10-03 |
| 2-04 | 10-04 |
| 2-05 | 10-05 |
| 2-06 | 10-06 |

| Forte Name | Complement |
|---|---|
| 3-01 | 9-01 |
| 3-02 | 9-02 *I* |
| 3-02 *I* | 9-02 |
| 3-03 | 9-03 *I* |
| 3-03 *I* | 9-03 |
| 3-04 | 9-04 *I* |
| 3-04 *I* | 9-04 |
| 3-05 | 9-05 *I* |
| 3-05 *I* | 9-05 |
| 3-06 | 9-06 |
| 3-07 | 9-07 *I* |
| 3-07 *I* | 9-07 |
| 3-08 | 9-08 *I* |
| 3-08 *I* | 9-08 |
| 3-09 | 9-09 |
| 3-10 | 9-10 |
| 3-11 | 9-11 *I* |
| 3-11 *I* | 9-11 |
| 3-12 | 9-12 |

| Forte Name | Complement |
|---|---|
| 9-01 | 3-01 |
| 9-02 | 3-02 *I* |
| 9-02 *I* | 3-02 |
| 9-03 | 3-03 *I* |
| 9-03 *I* | 3-03 |
| 9-04 | 3-04 *I* |
| 9-04 *I* | 3-04 |
| 9-05 | 3-05 *I* |
| 9-05 *I* | 3-05 |
| 9-06 | 3-06 |
| 9-07 | 3-07 *I* |
| 9-07 *I* | 3-07 |
| 9-08 | 3-08 *I* |
| 9-08 *I* | 3-08 |
| 9-09 | 3-09 |
| 9-10 | 3-10 |
| 9-11 | 3-11 *I* |
| 9-11 *I* | 3-11 |
| 9-12 | 3-12 |

| Forte Name | Complement | Forte Name | Complement |
|---|---|---|---|
| 4-01 | 8-01 | 8-01 | 4-01 |
| 4-02 | 8-02 *I* | 8-02 | 4-02 *I* |
| 4-02 *I* | 8-02 | 8-02 *I* | 4-02 |
| 4-03 | 8-03 | 8-03 | 4-03 |
| 4-04 | 8-04 *I* | 8-04 | 4-04 *I* |
| 4-04 *I* | 8-04 | 8-04 *I* | 4-04 |
| 4-05 | 8-05 *I* | 8-05 | 4-05 *I* |
| 4-05 *I* | 8-05 | 8-05 *I* | 4-05 |
| 4-06 | 8-06 | 8-06 | 4-06 |
| 4-07 | 8-07 | 8-07 | 4-07 |
| 4-08 | 8-08 | 8-08 | 4-08 |
| 4-09 | 8-09 | 8-09 | 4-09 |
| 4-10 | 8-10 | 8-10 | 4-10 |
| 4-11 | 8-11 *I* | 8-11 | 4-11 *I* |
| 4-11 *I* | 8-11 | 8-11 *I* | 4-11 |
| 4-12 | 8-12 | 8-12 | 4-12 |
| 4-12 *I* | 8-12 *I* | 8-12 *I* | 4-12 *I* |
| 4-13 | 8-13 *I* | 8-13 | 4-13 *I* |
| 4-13 *I* | 8-13 | 8-13 *I* | 4-13 |
| 4-14 | 8-14 | 8-14 | 4-14 |
| 4-14 *I* | 8-14 *I* | 8-14 *I* | 4-14 *I* |
| 4-15Z | 8-15Z *I* | 8-15Z | 4-15Z *I* |
| 4-15Z *I* | 8-15Z | 8-15Z *I* | 4-15Z |
| 4-16 | 8-16 *I* | 8-16 | 4-16 *I* |
| 4-16 *I* | 8-16 | 8-16 *I* | 4-16 |
| 4-17 | 8-17 | 8-17 | 4-17 |
| 4-18 | 8-18 *I* | 8-18 | 4-18 *I* |
| 4-18 *I* | 8-18 | 8-18 *I* | 4-18 |
| 4-19 | 8-19 *I* | 8-19 | 4-19 *I* |
| 4-19 *I* | 8-19 | 8-19 *I* | 4-19 |
| 4-20 | 8-20 | 8-20 | 4-20 |
| 4-21 | 8-21 | 8-21 | 4-21 |
| 4-22 | 8-22 *I* | 8-22 | 4-22 *I* |
| 4-22 *I* | 8-22 | 8-22 *I* | 4-22 |
| 4-23 | 8-23 | 8-23 | 4-23 |
| 4-24 | 8-24 | 8-24 | 4-24 |
| 4-25 | 8-25 | 8-25 | 4-25 |
| 4-26 | 8-26 | 8-26 | 4-26 |
| 4-27 | 8-27 *I* | 8-27 | 4-27 *I* |
| 4-27 *I* | 8-27 | 8-27 *I* | 4-27 |
| 4-28 | 8-28 | 8-28 | 4-28 |
| 4-29Z | 8-29Z *I* | 8-29Z | 4-29Z *I* |
| 4-29Z *I* | 8-29Z | 8-29Z *I* | 4-29Z |

| Forte Name | Complement |
| --- | --- |
| 5-01 | 7-01 |
| 5-02 | 7-02 *I* |
| 5-02 *I* | 7-02 |
| 5-03 | 7-03 *I* |
| 5-03 *I* | 7-03 |
| 5-04 | 7-04 *I* |
| 5-04 *I* | 7-04 |
| 5-05 | 7-05 *I* |
| 5-05 *I* | 7-05 |
| 5-06 | 7-06 *I* |
| 5-06 *I* | 7-06 |
| 5-07 | 7-07 *I* |
| 5-07 *I* | 7-07 |
| 5-08 | 7-08 |
| 5-09 | 7-09 *I* |
| 5-09 *I* | 7-09 |
| 5-10 | 7-10 *I* |
| 5-10 *I* | 7-10 |
| 5-11 | 7-11 |
| 5-11 *I* | 7-11 *I* |
| 5-12Z | 7-12Z |
| 5-13 | 7-13 *I* |
| 5-13 *I* | 7-13 |
| 5-14 | 7-14 *I* |
| 5-14 *I* | 7-14 |
| 5-15 | 7-15 |
| 5-16 | 7-16 *I* |
| 5-16 *I* | 7-16 |
| 5-17Z | 7-17Z |
| 5-18Z | 7-18Z *I* |
| 5-18Z *I* | 7-18Z |
| 5-19 | 7-19 *I* |
| 5-19 *I* | 7-19 |
| 5-20 | 7-20 *I* |
| 5-20 *I* | 7-20 |
| 5-21 | 7-21 *I* |
| 5-21 *I* | 7-21 |
| 5-22 | 7-22 |
| 5-23 | 7-23 *I* |
| 5-23 *I* | 7-23 |
| 5-24 | 7-24 *I* |
| 5-24 *I* | 7-24 |
| 5-25 | 7-25 *I* |
| 5-25 *I* | 7-25 |
| 5-26 | 7-26 |
| 5-26 *I* | 7-26 *I* |
| 5-27 | 7-27 *I* |
| 5-27 *I* | 7-27 |
| 5-28 | 7-28 |
| 5-28 *I* | 7-28 *I* |
| 5-29 | 7-29 *I* |
| 5-29 *I* | 7-29 |
| 5-30 | 7-30 *I* |
| 5-30 *I* | 7-30 |
| 5-31 | 7-31 *I* |
| 5-31 *I* | 7-31 |
| 5-32 | 7-32 *I* |
| 5-32 *I* | 7-32 |
| 5-33 | 7-33 |
| 5-34 | 7-34 |
| 5-35 | 7-35 |
| 5-36Z | 7-36Z *I* |
| 5-36Z *I* | 7-36Z |
| 5-37Z | 7-37Z |
| 5-38Z | 7-38Z *I* |
| 5-38Z *I* | 7-38Z |

| Forte Name | Complement |
| --- | --- |
| 7-01 | 5-01 |
| 7-02 | 5-02 *I* |
| 7-02 *I* | 5-02 |
| 7-03 | 5-03 *I* |
| 7-03 *I* | 5-03 |
| 7-04 | 5-04 *I* |
| 7-04 *I* | 5-04 |
| 7-05 | 5-05 *I* |
| 7-05 *I* | 5-05 |
| 7-06 | 5-06 *I* |
| 7-06 *I* | 5-06 |
| 7-07 | 5-07 *I* |
| 7-07 *I* | 5-07 |
| 7-08 | 5-08 |
| 7-09 | 5-09 *I* |
| 7-09 *I* | 5-09 |
| 7-10 | 5-10 *I* |
| 7-10 *I* | 5-10 |
| 7-11 | 5-11 |
| 7-11 *I* | 5-11 *I* |
| 7-12Z | 5-12Z |
| 7-13 | 5-13 *I* |
| 7-13 *I* | 5-13 |
| 7-14 | 5-14 *I* |
| 7-14 *I* | 5-14 |
| 7-15 | 5-15 |
| 7-16 | 5-16 *I* |
| 7-16 *I* | 5-16 |
| 7-17Z | 5-17Z |
| 7-18Z | 5-18Z *I* |
| 7-18Z *I* | 5-18Z |
| 7-19 | 5-19 *I* |
| 7-19 *I* | 5-19 |
| 7-20 | 5-20 *I* |
| 7-20 *I* | 5-20 |
| 7-21 | 5-21 *I* |
| 7-21 *I* | 5-21 |
| 7-22 | 5-22 |
| 7-23 | 5-23 *I* |
| 7-23 *I* | 5-23 |
| 7-24 | 5-24 *I* |
| 7-24 *I* | 5-24 |
| 7-25 | 5-25 *I* |
| 7-25 *I* | 5-25 |
| 7-26 | 5-26 |
| 7-26 *I* | 5-26 *I* |
| 7-27 | 5-27 *I* |
| 7-27 *I* | 5-27 |
| 7-28 | 5-28 |
| 7-28 *I* | 5-28 *I* |
| 7-29 | 5-29 *I* |
| 7-29 *I* | 5-29 |
| 7-30 | 5-30 *I* |
| 7-30 *I* | 5-30 |
| 7-31 | 5-31 *I* |
| 7-31 *I* | 5-31 |
| 7-32 | 5-32 *I* |
| 7-32 *I* | 5-32 |
| 7-33 | 5-33 |
| 7-34 | 5-34 |
| 7-35 | 5-35 |
| 7-36Z | 5-36Z *I* |
| 7-36Z *I* | 5-36Z |
| 7-37Z | 5-37Z |
| 7-38Z | 5-38Z *I* |
| 7-38Z *I* | 5-38Z |

| Forte Name | Complement |
|---|---|
| 6-01 | 6-01 |
| 6-02 | 6-02 *I* |
| 6-02 *I* | 6-02 |
| 6-03Z | 6-36Z *I* |
| 6-03Z *I* | 6-36Z |
| 6-04Z | 6-37Z |
| 6-05 | 6-05 *I* |
| 6-05 *I* | 6-05 |
| 6-06Z | 6-38Z |
| 6-07 | 6-07 |
| 6-08 | 6-08 |
| 6-09 | 6-09 *I* |
| 6-09 *I* | 6-09 |
| 6-10Z | 6-39Z |
| 6-10Z *I* | 6-39Z *I* |
| 6-11Z | 6-40Z *I* |
| 6-11Z *I* | 6-40Z |
| 6-12Z | 6-41Z *I* |
| 6-12Z *I* | 6-41Z |
| 6-13Z | 6-42Z |
| 6-14 | 6-14 |
| 6-14 *I* | 6-14 *I* |
| 6-15 | 6-15 *I* |
| 6-15 *I* | 6-15 |
| 6-16 | 6-16 *I* |
| 6-16 *I* | 6-16 |
| 6-17Z | 6-43Z *I* |
| 6-17Z *I* | 6-43Z |
| 6-18 | 6-18 *I* |
| 6-18 *I* | 6-18 |
| 6-19Z | 6-44Z *I* |
| 6-19Z *I* | 6-44Z |
| 6-20 | 6-20 |
| 6-21 | 6-21 *I* |
| 6-21 *I* | 6-21 |
| 6-22 | 6-22 *I* |
| 6-22 *I* | 6-22 |
| 6-23Z | 6-45Z |
| 6-24Z | 6-46Z *I* |
| 6-24Z *I* | 6-46Z |

| Forte Name | Complement |
|---|---|
| 6-25Z | 6-47Z *I* |
| 6-25Z *I* | 6-47Z |
| 6-26Z | 6-48Z |
| 6-27 | 6-27 *I* |
| 6-27 *I* | 6-27 |
| 6-28Z | 6-49Z |
| 6-29Z | 6-50Z |
| 6-30 | 6-30 *I* |
| 6-30 *I* | 6-30 |
| 6-31 | 6-31 *I* |
| 6-31 *I* | 6-31 |
| 6-32 | 6-32 |
| 6-33 | 6-33 *I* |
| 6-33 *I* | 6-33 |
| 6-34 | 6-34 *I* |
| 6-34 *I* | 6-34 |
| 6-35 | 6-35 |
| 6-36Z | 6-03Z *I* |
| 6-36Z *I* | 6-03Z |
| 6-37Z | 6-04Z |
| 6-38Z | 6-06Z |
| 6-39Z | 6-10Z |
| 6-39Z *I* | 6-10Z *I* |
| 6-40Z | 6-11Z *I* |
| 6-40Z *I* | 6-11Z |
| 6-41Z | 6-12Z *I* |
| 6-41Z *I* | 6-12Z |
| 6-42Z | 6-13Z |
| 6-43Z | 6-17Z *I* |
| 6-43Z *I* | 6-17Z |
| 6-44Z | 6-19Z *I* |
| 6-44Z *I* | 6-19Z |
| 6-45Z | 6-23Z |
| 6-46Z | 6-24Z *I* |
| 6-46Z *I* | 6-24Z |
| 6-47Z | 6-25Z *I* |
| 6-47Z *I* | 6-25Z |
| 6-48Z | 6-26Z |
| 6-49Z | 6-28Z |
| 6-50Z | 6-29Z |

## Appendix F.  List of Symbols

| | | | |
|---|---|---|---|
| $B$ | Barycenter | $\mathcal{O}_k(P)$ | $k^{\text{th}}$ orbit of $P$ |
| $B^n$ | Barycenter in $\mathcal{T}^n$ | $P$ | Generic chord |
| $\mathcal{C}$ | Complementation map | $\overline{P,(x_1,...,x_n)}$ | Orbit of a chord |
| $\delta(P,Q)$ | Metric distance between chords $P$ and $Q$ | $\overline{\overline{P,(x_1,...,x_n)}}$ | Class in $\mathcal{T}^n/\Phi^n$ |
| $\delta^{\Sigma}(\overline{P},\overline{Q})$ | Metric distance between orbits $\overline{P}$ and $\overline{Q}$ | $\|P\|$ | Evenness index |
| | | $P^C$ | Complement of chord $P$ |
| $\delta^{\Phi}(\overline{\overline{P}},\overline{\overline{Q}})$ | Metric distance between classes $\overline{\overline{P}}$ and $\overline{\overline{Q}}$ | $P_n$ | Passage matrix |
| | | $\pi_i$ | Reduction map |
| $\vec{e}_k$ | Unit vector in Euclidean space | $\widehat{\sigma_k}$ | Mesh of $k^{\text{th}}$ orbit |
| | | $\sigma$ | Permutation map |
| $\varepsilon$ | Primitive $n^{\text{th}}$ root of unity | $\Sigma^n$ | Cyclic group of permutations |
| $F$ | Fourier matrix | $\mathcal{T}^n$ | Interval space of $n$-note chords |
| $F^*$ | Conjugate transpose of $F$ | | |
| $f_i^j$ | Enrichment map | $\mathcal{T}^n/\Sigma^n$ | Quotient space over the cyclic permutation group |
| $\Phi^n$ | Group generated by cyclic permutations and inversions | $\mathcal{T}^n/\Phi^n$ | Quotient space over cyclic permutations and inversions |
| $I$ | Inversion map | $[x_1,x_2,x_3]$ | Pitch-class notation |
| $i(\overline{P})$ | $I$ – symmetry index | $(x_1,x_2,x_3)$ | Interval notation |
| Index $(P)$ | Evenness index | $[\![x]\!]$ | Greatest integer n such that $n \leq x$ |
| $Index_d(P)$ | Index of dissonance | $Y$ | $Y$- symmetry map |
| $\lambda$ | Eigenvalue | | |
| $M_H$ | Hankel matrix | | |
| $M_C$ | Circulant matrix | | |

# References

[1] Forte, Allen. *The Structure of Atonal Music*. New Haven: Yale University Press, 1973.

[2] Gray, R. M. *Toeplitz and Circulant Matrices: A review*. Foundations and Trends in Communications and Information Theory, Vol 2, Issue 3, pp 155-239, 2006

[3] Huron, D. and Sellmer, P. *Critical bands and the spelling of vertical sonorities*. Music Perception, Vol. 10, No.2, 1992

[4] Kameoka, A. and Kuriyagawa, M. *Consonance theory, part II: Consonance of complex tones and its computational method*. Journal of the Acoustical Society of America, Vol. 45, No. 6, 1969

[5] Morris, Robert D. *Class Notes for Atonal Music Theory*. Lebanon, New Hampshire: Frog Peek Music, 1991.

[6] Plomp, R. and Levelt, W.J.M. *Musical consonance and critical bandwidth*. Journal of the Acoustical Society of America, Vol. 38, 1965

[7] Rahn, John. *Basic Atonal Theory*. New York: Schirmer Books, 1980.

[8] Sommerville, D. M. Y. *An Introduction to the Geometry of n Dimensions*. New York: Dover, p. 124, 1958.

[9] Tenney, J. *A history of "consonance" and "dissonance"* New York: Gordon and Breach, 1988